ITALIAN *kitchen* GARDEN

First published in the United Kingdon in 2011 by
Pavilion Books
10 Southcombe Street
London, W14 0RA

An imprint of the Anova Books Company Ltd

Associate Publisher: Nina Sharman
Editors: Katie Deane, Helen Griffin and Barbara Dixon
Concept and jacket design by Georgina Hewitt
Layout: Rosamund Saunders

Photography: Ian Nolan
Recipe photography: Yuki Sugiura
See picture credits on page 176 for other photographs

ISBN 978-1-86205-910-8

A CIP catalogue record for this book is available from the British Library.

10 9 8 7 6 5 4 3 2 1

Reproduction by Rival Colour Ltd, UK
Printed and bound by 1010 Printing Ltd, China

www.anovabooks.com

Sarah Fraser

ITALIAN *kitchen* GARDEN

PAVILION

Contents

introduction

Italian food is considered to be among the best in the world. The reasons are simple: the food tastes wonderful, is healthy and easy to prepare. The best Italian food has a simplicity to it that leaves everyone yearning to reproduce that flavour at home.

So, what is the secret to great Italian food? Ask any Italian, from a qualified chef to the gnarled old *contadini* (peasant farmer) you still see in their vegetable plots today, and the answer is always the same: the best ingredients make the best food. You cannot compromise on ingredients, and any Italian will agree – the best of the best is what you grow at home.

But is it possible to grow super tomatoes and awesome-tasting aubergines outside the Mediterranean region? Absolutely. Growing Italian varieties of vegetables isn't mysterious at all. You simply need to understand that they are used to more sun and less harsh wind than in northern climes. In every other respect they are like all plants.

However, there are a few tricks that will help your growing vegetables think they are in sunny Italy and you need to understand and give each plant what it wants and needs. But, if you choose the right varieties and employ common sense, you will soon be well on your way to cooking with the best home-grown Italian ingredients. Ingredients that would pass muster in any Italian kitchen and will help you cook up that authentic Italian flavour in every meal.

Contadini are a dying race. In Italy it is viewed as very backwards to get your hands dirty; most young people aspire to living in a town apartment and never having to toil on the land like their *nonni* (grandparents) did. However, many *nonni* still do work a family vegetable plot, providing fresh produce for the whole family on a surprisingly small patch of land that has probably been in their family for generations.

When my family moved to Italy seven years ago the motive was to find a better lifestyle. In the UK we were a busy, two-income household. We were so preoccupied with 'aspiring' that we didn't realize we were falling apart. Life had become empty and meaningless. We worked, we played childcare tag, we dreamed of upgrading the car and the house and forgot that we had souls that needed nourishing too. Slowly it became clear that we were unfulfilled. But what to do about it? This was when fate made a grand appearance in our lives and dangled the juicy carrot of 'Casa del Sole' under our noses. We were trawling the internet for a holiday house to rent for a week when we came across the farm for sale. Just like thousands of other people do, we played a fantasy game of 'what if?'. Should we stay in the rat race and follow our path with noses to the grindstone? Or should we break out of the box and try something new? We chose the new life; not because we are brave but because we needed a change and an adventure was calling to us. Casa del Sole was an idyllic, if decrepit, farmhouse in the Tuscan hills. Gardens, fruit trees, olive trees and woodland all combined to weave a dream of country living. I had big ideas about self-sufficiency (fuelled by the fact that we didn't have any money to live any other way) but little knowledge of the skills needed to exist off the land.

A ogni uccello il suo nido è bello

There's no place like home.

I was lucky when I started my kitchen garden because our neighbours not only were friendly and helpful with all things to do with gardening, from giving me seedlings to telling me where to buy the best implements, but their grandfather had owned our house and the next two generations had been born and raised in it. This meant that they were the ideal people to ask about siting for new garden plots and what grew best where.

One of the key objectives of our new lifestyle was to produce as much of our own food as possible. I quickly learned that I would rather be vegetarian than eat my darling chickens. From that moment of realization that I could no more eat my pet goose than eat one of the kids, I threw myself into making the best kitchen garden possible on our plot… it was the only way we were going to get a good meal!

I come from a family of gardeners. And being from Yorkshire, they were a little obsessed with getting 'owt for nowt' so grew vegetables rather than flowers. Some of my earliest memories are of the awe-inspiring mystery surrounding planting seeds in the ground and watching the green sprouts emerge and turn slowly into something that would end up on my plate. From then on I was hooked. Whenever we moved house as a family the first job was to dig a vegetable patch (and make sure it was bigger than the neighbours'). Imagine my horror when I married my husband, who only had a concrete back yard of 5x5m (16x16ft). Within the week I had a cascade planter of strawberries, a dwarf apple tree with carrots seeded under it, a barrel of potatoes and countless pots of herbs. But I'm no trained gardener, and moving to a completely different climate meant my learning curve was steep. Yet my early experience has lent me an appreciation of how to explain gardening in simple and practical terms. Every year I learn so much and enjoy my garden to such an extent that I have become evangelical about gardening. I want everyone to have a go at it and experience how much their lives can be improved by taking care of a simple vegetable plot.

Non tutti quelli che hanno lettere sono savi
Not all those who are learned are wise.

Whether you have a tiny balcony or an acre, it really doesn't matter. It is the contact with the earth, the seasons and those elemental forces that influence gardening, that are the key to a happier, healthier life. Of course, you get a good dinner out of it too.

Experience has revealed to me that cooking with food from your kitchen garden is absolutely not like cooking with food from the supermarket. There are specialist skills that the kitchen garden cook needs to learn and appreciate. This is why the second half of this book is devoted to explaining the best traditional and often ancient methods and suggesting delicious recipes especially for the gardening cook.

There is no great mystery surrounding gardening; green fingers are not simply a lucky attribute inherited from some long lost ancestor. A good gardener is someone who has a rudimentary understanding of how plants grow, who understands something about the cycle of nature, and who is sympathetic and dedicated to nurturing new life.

When we arrived in Italy one of our lovely neighbours came to the house to welcome us and invite us around for dinner. I asked what she would be cooking and if I could bring anything. She replied that dinner depended on what was ripe in the garden that day and that I should bring nothing but the opportunity for her to practise her English. The idea that dinner was entirely dependant on nature rather than a trip to the supermarket was a complete revelation.

Ogni regola ha un' eccezione
There is an exception to every rule.

Gardening isn't complicated or difficult. Anyone can do it by understanding a few basic principles:

• Plants and seeds want to grow: it is their destiny. All you have to do is give them a fighting chance against weeds and drought.
• Seeds turn into plants by being watered, well-drained, warm, and by being given sunlight once they poke their heads above ground level.
• Plants only fail if there is a problem. They are not fickle. If they do not grow correctly or well then this is their only way of telling us that there is a problem that we need to resolve.
• Problem solving is detective work with plants – since they cannot tell you what the problem is. We need to be observant and to know a little about pests, diseases, light and shade requirements and soil nutrients. And we should have a good reference book on hand.

• You reap what you sow. It's absolutely true that the amount of work you put into a garden is directly paid back in satisfaction and yield.

• Soil quality is paramount. Thus, a good working knowledge of your own soil and the principles of composting is an essential, if somewhat smelly, lesson.

EQUIPMENT

Beyond your ordinary gardening equipment, you will need some gardeners' fleece, a few empty old plastic bottles and a greenhouse or sunny window with a wide sill.

LEFT: Adding ash to your garden puts back all the minerals from the wood into your soil.

SITE

Take a good look at your site. You need to be realistic about how much full sun your site is receiving at different times of the day. Generally, a 'sunny plot' will receive a minimum of six hours sunshine a day. Usually, plants that produce fruit will need the most sunshine to ripen it. This would include tomatoes, peppers and cucumbers. If you have a shady plot then do what you can to make as much sunlight as possible (I have heard of people putting mirrors on back walls to maximize the rate of ripening for their tomatoes). If you can't prune or lop the trees that are creating shade you will need to choose varieties of vegetable that require less sun. In a dappled, partly shaded area you can grow 'frutti di bosco' (currants, strawberries, gooseberries); also some vegetables, such as broccoli, leeks and spinach, will tolerate less sun. In an area of total shade the only thing you can plant that might yield anything is Alpine strawberries.

If your site is boggy then don't even think about planting Italian vegetable varieties. You will first need to irrigate or create raised beds that generally drain freely and are most suited to growing vegetables.

SOIL

The soil in Italy is as diverse as in every other country, but with soil there is an international rule: well-fertilized, well-drained soil produces the best crops. The only real difference is that in Italy there is generally less rain and more sun, so northerly gardens need to have free-draining soil to accommodate Italian native vegetables and you should try to create sun-traps and warm shelters for your Mediterranean veggies.

La gente in case di vetro non dovrebbe gettare le pietre
People in glass houses should not throw stones.

Once you start growing and cooking your own dinners you enter a bright new world where food is both respected and admired. The garden nourishes your soul and the vegetables nourish your body – perfect harmony. Of course, sometimes you might hear the odd phrase such as, 'Not courgette again!', but at this point I would advise you to make courgette cake and watch how quickly it gets scoffed.

One of my favourite times of year is early springtime when I plan what I am going to plant and where. This is when you will find me scouting through seed catalogues and browsing our local plant nursery's market stall, and at home with a pencil and paper mapping out my beds.

When I first started gardening in Italy I wasn't sure about how much each plant might yield and consequently had a few gluts of vegetables that no one liked. (I can't face a courgette again after eating it prepared in every way imaginable.)

THE RULES

RULE NUMBER ONE: don't plant something that you know no one in your family likes, even if it is very easy to grow and someone gave you the seedlings. It's just not worth the effort.

RULE NUMBER TWO: know your average yield (which I have indicated for every plant listed) so you can plant the correct amount for your desired harvest – and then plant a few extra, just in case.

RULE NUMBER THREE: there are no more rules – enjoy yourself and plant whatever you want!

I have listed all the fruits and vegetables necessary for a gourmet gardener to make a well-balanced and varied kitchen garden. However, once you get going you will find that you want to experiment with other varieties and species – let your imagination and your taste buds go wild.

Happy digging and baking!

GETTING Started

growing the italian way

WHY GROW ITALIAN FOOD?

Growing your own Italian varieties of fruit, vegetables and herbs is the best way to ensure that you have the freshest, tastiest ingredients for your kitchen. If you love authentic Italian food you will know that it is important also to have the specific varieties that are grown in Italy – any old tomato is not acceptable to the Italian chef!

Growing Italian varieties is also a challenge because some you may not have heard of and you may need to modify your gardening techniques to grow many Italian vegetables in Northern climes. Don't let this put you off. Ultimately it is no more difficult to grow Italian fruit and vegetables than it is to grow traditional stock. You just need to know where to start.

CAN YOU GROW ITALIAN VEGETABLES?

CLIMATE

The further north you go the colder and wetter the climate becomes. In general, Italian plants like the sun, dislike too much rain and aren't used to the wind. There are ways to compensate for this. Firstly, you need to make sure that your site is well drained. Secondly, that it is sheltered, and thirdly that the most sensitive plants are positioned in the sunniest, south-facing positions. Be very aware of the light situation in your garden. Shade in Italy is sought after because mid- to late summer is so hot that the sun can literally scorch a plant's leaves in an afternoon. Further north, the sun is not so strong in summer and full exposure is necessary to ripen certain fruits and vegetables.

PESTS

Slugs and snails are not a major hazard in Italy from late spring onwards, as the ground is so dry that they can't slime between plants. Further north these little pests are the destroyers of gardens throughout summer and need some

ferocious attention. The problem for some Italian varieties is that they don't have enough foliage to survive a slug attack. For example, Tuscan Cavolo Nero (black cabbage) has long strappy leaves and slugs can eat almost a whole young plant in one sitting. There are various simple and organic methods to discourage slug attacks. Circle your plants or beds with a wide border of sand so they can't get over it (it sticks to their slimy underside and they can't progress). As double protection, sprinkle used damp coffee grounds around the base of your plants and among seedlings. This works in the same way as the sand but has the added bonus of tasting really nasty to slugs. Most other pests are the same in the UK and in Italy. Just as with all seed manufacture, Italian seeds have been bred as much as possible to be resistant to pests. So, if you choose the modern varieties you are slightly less likely to have a pest problem than if you are growing a 'heritage' variety.

Quando il gatto non c'è il topo balla

When the cat's away, the mice will dance – and eat your seeds!

SOURCING

Sourcing authentic Italian seeds and plants can be a challenge, as the average garden centre doesn't carry a wide variety of Italian plantlets and seeds beyond tomatoes and peppers. You may find it necessary to mail order from specialist seed companies such as The Italian Seed Company (www.italianseedcompany. co.uk) or Seeds of Italy (www.seedsofitaly.com). The internet is a wondrous place to find all sorts of traditional Italian seeds. One word of advice, though – try to buy from local suppliers, especially if they produce the seeds in your own climate (i.e. not shipped in from Italy). This is because seeds are living organisms, grown in a particular environment and they sometimes find it hard to adapt to a new climate. Thus, imported seeds always have a lower germination rate than home-produced ones. This isn't scientifically proven, but it's what the locals say and it certainly proved true for my English imported seeds. But if you can only find imported, go ahead and plant them, but plant more and be very kind to them. Second generation seeds, however, seem to do just fine.

SMALLER SCALES

Even if you have just a patio garden, balcony or a window box, you can grow something. You may wonder if it is worthwhile growing your own herbs or having a tub of tomatoes taking up space on your windowsill. My answer would (of course) be 'Yes': because growing things gives you a sense of achievement and freedom; because you can feel good about yourself for having a hobby that is nourishing to your physical wellbeing; because it's something you can do that will benefit the environment for a change. And finally, just because you can.

I look on growing my own food in the same way I feel about my children. They are hard work; they are challenging; they are downright annoying sometimes. If you calculated the cost and time input in raising them you'd certainly decide not to have them on economic grounds. But I wouldn't be without them and every day they teach me something new and make me a better person.

One year early on in my gardening career, our friend Luciano came around with his car boot full of young plants. He had sown way too many of everything and he thought I might like some. I was grateful (there is always room for more plants in my garden) and set about hefting the crates into the veg patch. Unfortunately, Luciano was in a rush and shot off in his van before I had a chance to ask him what everything was. I could identify half of the plants but melons, cucumbers, butternut squash and courgette plants all look alike when young. I ended up trying to train a courgette up a trellis because I thought it was a cucumber and planting melons in with my squash. Which was fine really – I just felt a bit daft when people asked if there was a scientific reason for my interplanting the crops.

TIME AND MONEY

Gardening should not cost you too much in terms of money as long as you are willing to recycle and be imaginative. You can plant in the most bizarre of recycled containers; yoghurt pots can be used to plant seedlings, old wood can be used to create raised beds, recycled plastic bags can be made into a mini polytunnel, old windows can be made into a cold-frame. Growing your own vegetables will certainly save you money in the long run.

Time expenditure is another thing entirely. Gardening is a high maintenance pastime. Notwithstanding any extra initial set-up time for your garden, you will need to spend an average of ten minutes per metre square per week. This means that scale is an issue. Be practical in your estimation of how much time you have available to spend on your garden when deciding how big to make it. Remember also to factor in how much time you want to spend gardening per week. Personally, I find escaping to the garden therapeutic. I also enjoy teaching my children about growing things and they love to help pick the produce (although they are not so enthusiastic about weeding).

Time spent on your garden is also seasonally intensive. Although I have just given an average of ten minutes per msq, this is slightly misleading because, in spring when the weeds are growing thick and fast, you will be spending more time than that weeding. In winter though, when there is not so much to do, you will be able to sit back and relax.

HEALTH

If you have never done any gardening before then I would recommend you start slowly. Gardeners often complain of bad backs because of all the digging and bending over to weed. The truth of the matter is that if you are using a good technique and your posture is good then you should not injure yourself at all (although you will tire yourself out to the point of collapse quite regularly).

Gardening is not heavy physical work but regular gardening will help keep you fit and active. It is a great way to lose weight and tone muscle; pace yourself though. If you feel a twinge in your back whilst you are weeding then move on and do something else – the weeding will still be there when you return to it. Do no single task for longer than 15 minutes at a time. Before starting to garden always employ a few simple stretches to loosen your muscles (never mind that you might look a bit daft – it's better than ending up injured and it will improve your posture and muscle tone).

RIGHT: A good herb garden is essential for any serious cook. Fresh herbs are a world away from dried ones.

setting up

Chi ben comincia è a metà dell'opera
A good start is half the battle.

SITE DESIGN AND PREPARATION

It really doesn't matter what size plot you have to give over to your kitchen garden. In fact, you can have a good crop of various veggies just from window ledge gardening so don't let questions of scale put you off.

When I first plotted my garden I thought that 200msq (an average sized back garden for a suburban semi-detached house) was going to be insufficient in the long run but decided to 'start small'. In fact, I have found that this size is perfectly adequate to feed my family and I only use more space for large crops such as sunflowers and sweetcorn that is predominantly grown to feed the chickens and rabbit through winter.

WORK WITH WHAT YOU HAVE

Once you have decided to create a kitchen garden, the first rule is to optimize what you have. If you have a large plot you need to work out which part of it to use, which part is in the sun longest, where the water is and what the drainage is like.

GARDEN REQUISITES

OPTIMUM SUNLIGHT

Sunlight is necessary to make plants grow (they convert sunlight, by a nifty process called photosynthesis, into energy that they then use to grow and to reproduce). High levels of sunlight also mean warmth, so a site with plenty of sun warms up quicker in spring and the plants grow earlier, which gives them a longer growing season. In turn, this means bigger, more fruitful plants.

The best kitchen gardens have full sun for as much of the day as is possible. If you have any say in the matter, choose a south-facing slope. If you find your garden is shaded by trees or bushes you may want to do some pruning to let in adequate light.

If you are growing on a balcony or in window boxes, by definition they are generally pretty light so there should be no problem. However, if you have the odd nook that is shady, the best thing to do is plant something that grows happily in the shade. In shady corners where your garden is so small that you need to utilise every inch, you can try mounting mirrors on a back wall to improve your light – it really does work.

WATER

Plants need water on a constant basis. Even in wet climates you will need a ready supply of water for your garden.

Your water will probably be on tap and near to your garden but it always helps to recycle rainwater to save the environment and your pocket. A couple of well-placed water barrels to catch the rainwater off your roof is an excellent idea. Many plants thrive on rainwater and hate tapwater because of the added chemicals. My Grandfather swore that his success with tomatoes was down to watering them with the rainwater off the garage roof.

Essential for container and raised bed gardening is a good watering system. If you're keen then I recommend that you install an irrigation system, preferably one that you can put on a timer so that you can go away and not have to worry that your plants will dry out. Small-scale irrigation is not nearly as complicated as it sounds. You can buy what is effectively a hosepipe that has little leader pipes off it with spikes that you press into the soil and the water leaks out directly onto the pot. You simply connect the hose to your outdoor tap and position the spikes. It's cheap, easy to figure out and well worth it in terms of time saved. The addition of a timer on the tap (again, easily available at garden centres) makes for almost maintenance-free gardening.

DRAINAGE

Rather irritatingly, plants may need lots of water but they hate to be waterlogged and so good drainage is essential.

While it is possible to improve drainage, it is quite difficult and time-consuming, so choosing a site with a slight slope is a good idea. Never choose a boggy site for your garden, as putting in drainage ditches is real hard work and nothing will grow in waterlogged ground apart from watercress.

If you find that your garden is badly drained then you can take measures to combat this – but they will be extensive. In a small garden, raised beds are usually the most economical way of improving a badly drained site.

Pots and containers should drain well. Make sure they have a few rocks or

LEFT: Raised beds, and borders
between your planting areas,
help you to keep your garden tidy
and ordered.

lumps of polystyrene in the bottom so that the drainage hole doesn't get bunged
up with soil. Roots will rot if stood in water.

SOIL QUALITY

Soil quality is not as mystical as it sounds. It's either good for growing stuff or it
needs improving. Bad soil is no excuse to give up before you have begun.

Unless you can see anything obviously wrong with your soil (very heavy
clay or overly sandy) then don't worry too much about your soil to begin with.
Eventually, you might find you are having problems growing certain crops – this
is the time to start testing your soil.

If growing in pots and containers you'll probably be using bought-in
compost but don't be afraid of using topsoil, if you can pilfer some, and mixing
it with a generous helping of manure. Bought compost has the advantage of

being sterile, so it contains no other seeds or viruses but good soil and a bit of manure will mean that you don't have to worry about fertilizers.

WIND

A windy plot is always going to present you with problems, from physical damage to crops to the fact that it's always colder in the wind. You can put in windbreaks for plants but it's best if you can start off with a sheltered spot if possible. Generally, your back garden will benefit from urban shelter and so you need do nothing. If this is not the case you can erect fences, plant bushes or hedges and organize your garden so the taller plants shelter the smaller ones.

Sometimes balconies and windows are subject to quite a bit of wind and it's difficult to make any shelter. Under these circumstances you are advised to stick to low-growing plants that are sturdy. For example, choose a bush variety of tomato and don't try to grow sunflowers!

The perfect kitchen garden has:
- Optimum sunlight
- Water
- Drainage
- Good soil
- Shelter

window boxes
& containers

Many vegetables and fruits grow very happily in containers, so don't worry if you are short on garden. My first ever garden was 4msq (4sq.yd) and I happily grew an apple tree, loads of herbs and strawberries in it. In fact, container gardening is sometimes easier than large plot gardening because the weeding is kept to a minimum.

SUNLIGHT
By definition, windows and balconies are generally pretty light. If you have the odd nook that is shady, the best thing to do is plant something that grows happily in the shade.

WIND
Often balconies and windows are subject to quite a bit of wind and it's sometimes difficult to make any shelter. Under these circumstances you are advised to stick to low growing plants that are sturdy.

WATER
Essential for container gardening is a good watering system. Pots and containers dry out very quickly when there is little rain and many plants can die within a couple of days of their roots drying out. Certainly most plants will not recover fully from a bout of arid dryness.

Small-scale irrigation is not costly and means that you can go away for the weekend without having to ask your neighbours to look after your plants.

DRAINAGE
Pots and containers should drain well. If they don't then the roots will rot and the plant die. Check regularly that your pots are not marshy and if you suspect they are then have a look to see if the drainage holes are blocked – this can easily happen with small stones, debris and roots.

SOIL
Container compost is generally higher in fibre than normal topsoil so that it can absorb more water. However, it is always good to introduce a little natural topsoil if you can to the compost because worms and minerals are vital to the healthy growth of vegetables.

I have a lovely Italian friend called Maida who is a great gardener but is also very particular about her hands and nails. She is forever pointing out to me that I need not sport the 'peasant look' with dirt ingrained into fingers and nails that are always slightly mucky: if only I wore a pair of latex gloves. While the idea is a good one, I'm not organized enough to manage this yet myself.

SOIL

Soil is the foundation of any garden. A good soil is dark, rich in compost (for nutrients), has a small amount of sand for drainage and isn't too claybound. However, any problem soil can be corrected by the addition of organic matter.

TYPES OF SOIL

Loamy soil: This is the perfect soil for growing vegetables. It has a roughly even sand, silt and clay content with high levels of organic matter (which absorbs water and nutrients, and prevents it from drying out too much). It feels quite gritty (which allows for good drainage when it is wet) and is easy to dig.

Clay soil: Clay is claggy and doesn't drain well. When dry it is cement-like and cracks, so that water isn't easily absorbed. To correct a clay soil dig in at least 30cm (12in) depth of manure every autumn on top of the normal manuring/compost distribution throughout the year (see 'Muck', p.33). Your soil should be 100 per cent better after two years, even in the worst of clay areas, and transformed into a good growing medium. In an emergency, you can dig in this amount of compost/manure at any time of year but make sure that it is well-rotted so you don't get lots of rogue weed seeds germinating in your new plot.

Sandy soil: Soil that is overly sandy is usually a rust colour, gritty to touch, dries out to dust and has little organic matter in it with which to absorb and store valuable nutrients and moisture. Organic matter needs to be added in the same way as the correction method for clay soil, above.

Silty soil: Silty soil is also excellent for growing. It tends to be made up of minerals and very fine organic particles. It retains nutrients well, drains well and because of its very soft consistency is easy to work with.

Peaty soil: Peaty soil is full of organic material and can be acid. It will need correction by adding sand, silt and well-rotted organic matter until the balance is returned. Watch out that this soil doesn't retain too much water – if it gets very boggy it will rot the roots of plants.

Chalky soil: Chalky soils are usually full of stones and quite alkaline. They drain too freely and don't have much in the way of nutrients. This type of soil can be improved by adding a great deal of well-rotted vegetative matter. Preferably a barrow of muck per msq, deeply dug twice a year.

SOIL ACIDITY

Your soil can be naturally acid, neutral or alkaline, and this is measured in pH on a scale of 1–13. Most soils are between pH6 and 8 and this suits most plants. However, some plants are more fussy and prefer a particular pH (brassicas like alkaline soil, for example). You can buy a simple test from garden centres that will give you the pH of your soil (check your whole garden since pH can change from bed to bed). Once you have established what the imbalance is, you can address it by adding sulphur if the soil needs to be more acid, or lime if it needs to be more alkali. But don't worry too much because if you are following a good manure and composting routine your soil should be fine. However, if a crop persistently doesn't do well then check the soil acidity for irregularities.

SOIL NUTRIENTS

Good soil holds all the nutrients a plant needs to grow and produce fruit. However, when we create a garden we are artificially overloading that patch of earth and making unnatural demands on the nutrient levels. This is why we need to concentrate as much on what we put into the soil as what we grow out of it.

LEFT: Baby carrots need to be sown in their final position, as you cannot move the seedlings. They like light, sandy soil that has been well worked.

A serious manuring/fertilizing regime must be maintained throughout the year (see 'Muck', below). If you find that a crop is failing and you have checked the soil for acidity and things seem OK then it will be necessary to check for nutrient deficiency. Again, you can buy a kit at garden centres and correct the problem by adding the relevant element. I find that my tomatoes often lack calcium and so get rotten bottoms and therefore I make sure I add lime to the soil in autumn and all the shells of any seafood we eat get crushed up and dug in. Also, hair from hairbrushes and all nail clippings go in there too. Broccoli tends to suck up nutrients so needs a particularly rigorous manuring.

Solo chi non fa, non falla

Only those who do nothing, make no mistakes.

MUCK (AND THE IMPORTANCE OF IT)

If you read the section 'Organic Matters' (p.38) you will realize that I am an organic gardener through and through. As such there are some simple principles that one needs to understand when looking after your garden…

There are only a few basic chemicals that make up everything in the world. These components are put together to create organic matter. When that matter ceases to live, it decomposes into its original basic components. This is the perfect circle of life. Nature intended everything that lives to die and go back into the land so that it can be regenerated into another form of life.

When we remove this dead matter and put it somewhere else to decompose then the ground it has grown from is depleted. This ground will need restocking with the nutrients that have been removed from the natural cycle.

When we burn the dead matter the original components will still go into the ground – although the carbon release is much faster than decomposition and can cause a threat to the environment, so natural decomposition is always best if possible.

Hence, when you create a vegetable garden you are building a place where you know that you are removing, on a constant basis, the goodness from the soil. Your plants will grow, they will fruit, you will eat the fruit and the waste from you is flushed away. Although at our farm we have our own 'off grid' waste system (which composts all our human manure and releases it slowly and safely onto our land), this does not release into the vegetable garden, so I still need to pay close attention to ensuring that the balance of nutrients high.

In the old days, before a national sewage system, every house had its own 'midden' or muckheap. It must have stunk to high heaven but I bet their gardens were healthy. I am not suggesting that we return to this arcane system but I do think it's important we understand that what we take out we must replace.

So, how do we make sure our garden soil is in top condition? If you're lazy you can use chemicals that have been prepared for you to sprinkle on your garden with great ease. These are most often nitrate fertilizers. The problem with these industrial chemicals is that they do not improve the texture of your soil (just look at the section on growing carrots to see how important that is) and they get washed out of the soil by rain so easily that sometimes you may as well have saved your time and money. You may also want to spare a thought for where they get washed away to… the local water courses are flushed with nitrate fertilizer, which imbalances the ecosystems and wreaks all sorts of havoc on local wildlife.

COMPOST

Fortunately, there is a much safer way to feed your garden: build a good compost heap. It appeals to a Yorkshire lass like me because composting is getting something from nothing. A well-fed compost heap will take all your vegetation waste; all the stuff from your kitchen and your garden that you don't need will be converted into rich food for your garden. You can even compost weeds.

HOW TO MAKE A COMPOST HEAP: Unless you have a very small balcony you can make compost. For a mini version you can buy a beehive-style compost bin from your local garden centre (or save yourself a few quid and use a normal black plastic bin and drill a lot of holes in the bottom and sides).

Technically, a compost heap can be just that – a heap into which you pile all your organic waste. However, if you do it this way it normally spreads out and is not only unsightly but it also loses a lot of the important heat (the more heat build-up the more chance that weed seeds and any naughty virus spores will be deactivated). So, I would advise one of two options. If you are a neat person and like order you should build your compost with three slatted wooden sides and a firm (concrete is best) base. Next to this you would build a second compost bay so that you can have one compost heap on the go whilst the other is composting or being used up. I am not neat, and persuading Richard, my husband, to devote precious time or money to build a compost heap containment system isn't ever going to happen. So I use chicken wire. I bang four stakes into the bare ground to create a three-sided bay, then put the chicken wire around them and that's it. It looks slightly better than a 'heap' and it works better too because the heat is contained and it's easy to scoop the compost out when you are ready to use it.

WHAT CAN GO INTO A COMPOST HEAP? You can feed the compost heap with everything of vegetative origin that is going spare in your garden and kitchen. All your peelings, weeds, potato tops, etc., go into it. Newspapers (ripped up so that they decompose easier), grass cuttings, leaves… everything. Do not put animal products on your compost heap. Although they will rot down eventually, they will stink to high heaven and attract all unwanted scavengers.

ABOVE: Cabbage plants start off small, but grow very large, so they need plenty of room once they are transplanted to the bed.

HOW DOES IT WORK? The compost heap will then begin to rot. This process creates heat that hopefully kills the weed seeds. Don't underestimate how hot a muckheap can get. When we first moved to our farm we did not have any running hot water. We set up a wacky system of black hosepipe that ran through a compost heap full of grass clippings and then into our outdoor shower. The water reached such temperatures that they could scald you! A good compost heap is a steaming compost heap. If you find that your compost isn't rotting away as you would hope, you need to feed it. There is no delicate way to put this… pee on it. Urine speeds up the decomposition process of plant matter. It's the combination of water and chemicals that have the magic effect so don't be shy, get the potty out!

WHAT DO I DO WITH IT? After about three months in summer and perhaps six months in winter (reliant on temperature) your compost heap is ready to be dug into your garden. Depending on what you put into it, it should be a rich, dark, sometimes very thick, mass of compost, enormously nourishing for your garden. Pile it on as thickly as you are able. It can be added at any point in the year and will always give the soil a good boost. Don't put it straight up against plant stalks or stems – it can be a bit rich and there is the danger that it will still be rotting and releasing heat – but leave a few inches around the base of plants.

You can use this compost as a soil improver and dig it in, or, if your plants need a feed, just pile it on top of the soil and let the rain and the worms do the hard work.

MANURE

Manure and compost are different end products of the same decomposition process. Whilst composting is natural decomposition, the manure process is speedier and more efficient. Basically, manure is matter that has been through the gut of any animal (including humans). The process of digestion (depending on the species of animal) helps to release nutrients in the eaten fodder that then fertilize the land in the form of dung. The different animal species have different digestive mechanisms and so create dung with varying qualities. For example:

A chicken will eat grass and its dung will be high in nitrogen but low in vegetative matter. It will benefit little from further rotting. This is a 'high potency' plant food that can be used in the garden but is so strong it can 'burn' stalks and leaves, so always use either well-dug in or scattered on top of the soil away from stalks.

A horse will eat the same grass as a chicken but its dung will have fewer nitrates (a horse's digestion is designed to cope with bulk feed and is not as effective as a chicken's) and contain much more fibre, which is good for the condition and drainage of the soil. Horse manure is traditionally used 'well-rotted' because this ensures that, if it is used directly on the soil as fertilizer, the decomposition process is finished and no more heat will be released (plants can literally be scalded by decomposing manure). The other reason manure should be well-rotted is that this denatures (kills off) many of the seeds that pass through the horse's gut. Most of these are grass and weed seeds – just about the last thing you want to be introducing into your vegetable garden. It takes about eight weeks in summer for pure horse manure to rot – if it is mixed with straw add four weeks, if mixed with shavings add three weeks.

Manure is readily available from your local stables (if it's free, then you normally bag your own – better because you can search around for the well-rotted stuff). Any poo is manure, so if you know someone with pigs wanting to get rid of some manure then jump at the chance. Pig manure is high in nutrients because pigs are omnivores and have normally been eating a huge range of food: all the better for your garden. I like to mix my chicken muck in with my normal compost to give it extra oomph.

If your manure is well-rotted then you can just spread it around wherever it is needed. In summer you can put it in between your rows of plants – the

Owning horses is great – we swap our manure for wine in late autumn with our neighbour who has vines and in deep winter we save a fortune on fertilizer when every one of our 400 olive trees needs a good dose of extra nutrients.

weather and creepy crawlies will work it into the ground for you. If you are digging, put a few shovelfuls in at the same time; it will improve not only the nutrient level but also the texture of your soil. It's sensible to make sure that you have plenty of manure in spring and in autumn when you are changing your beds and preparing your soil. Don't forget the 'what you take out you must replace' principle, so never be stingy with manure.

If you find that you have unrotted manure, don't ignore it. The system I use is this: in autumn I cover all the beds that are not in use with unrotted horse manure to a depth of about 30cm (12in). It doesn't matter if it starts decomposing then and there because there is nothing in the beds for it to burn. Over the course of a couple of months this 'raw' manure will sink naturally into your ground. The best news is that you don't need to dig it in and it smothers the winter weeds through to spring. Once the weed seeds start to sprout it is almost time for planting and you should just give it a quick fork over (which is easier and quicker than digging), which will eliminate those pesky weeds. Missing out on a lot of hard digging seems to be a bit of a cheat really, but I am not complaining. One thing to note is that some vegetables (notably carrots) don't like a freshly manured bed, so leave at least one bed unmanured to keep them happy.

ORGANIC MATTERS

I believe wholeheartedly in organic gardening. I do not believe that any other type of gardening is necessary or helpful to the environment. For me, the only (and most successful) type of gardening is that which works with nature, using nature's own techniques to get the best out of your plot.

WHAT IS WRONG WITH USING PESTICIDES AND FERTILIZERS?

Firstly, these substances are artificially created. This means that you are right at the end of a production line and are paying extra to make someone else

In many parts of rural Italy people take their water from a stream. At Casa del Sole we use this system and are entirely at the mercy of our neighbours up the hill. If they decide to use pesticides these are washed directly into our tapwater. We are fortunate that we know our neighbours well and we are sure that they garden organically like we do.

In Italy there are greater restrictions on the use of pesticides than in the UK. The penalties for contaminating water sources, especially those that are registered as water sources for houses, are really stiff.

LEFT: You can recycle everything
in your garden. All you need to
remember is to make plenty of
drainage holes in the bottom of
recycled containers.

rich. Whilst this is fine if the result is something you can't do yourself, making
compost or fertilizer is so simple that you really should save yourself the time
and money and have a go. Be self-sufficient in as many ways as possible.

Secondly, the chemicals used in fertilizer are raw and have been produced
in a way that then releases other chemicals into our environment. When these
unnatural fertilizers are put on your garden, they do not act in a slow-release,
'natural' way but zap the plant and soil with high doses of nutrient or chemical.
The unused nitrate/chemicals then leach into waterways, altering natural
habitats forever.

Thirdly, there is conclusive evidence to prove that the chemicals that are
used on our fruit, vegetables and cereals to make a bigger, healthier-looking crop
are often absorbed into the food, consumed by us and then released into our
bodies to ill effect. It is a scandal that food should be treated in a way that may
actually make it harmful to us.

Finally, my heartfelt philosophy is to 'live lightly on the Earth'. This means
that I try to make as little irreversible impact on the planet as possible; I recycle and
use natural materials wherever possible. Composting, creating your own fertilizer
and using manure to help your plants grow is Nature's own form of recycling.

eco gardening

SLUGS

Slugs are a real problem in cooler, wetter climates (and in Italy in winter). Protect seedlings and young plants by cutting off the bottom of clear plastic bottles and placing them over the seedlings until they are big enough not to be ravaged by a slug attack.

• Put used coffee grounds (ask your local café for their used grounds) around your plants in a solid barrier – slugs hate them.

• Sink the bottoms of those cut-off bottles in the ground in your vegetable beds. Fill them with beer and cover with a tile so there is just a couple of centimetres between the rim of the container and the tile roof. Slugs love beer and covered moist hideaways, so you'll catch loads.

• Make sure that, in summer when it is dry, you only water around the base of the plant: slugs can't slither about on dry ground, so don't make it easy for them by getting the whole place damp.

• Ducks are absolutely brilliant at getting rid of your slugs. But don't let them in your garden until your plants are robust enough to survive a misplaced duck foot.

SEEDLINGS

Seedlings grow well in recycled yoghurt pots (remember to make three holes in the bottom so they drain well) and cardboard egg cartons. I use recycled plastic containers with lids from the supermarket to make mini greenhouses for my seedlings. Always remember that if you use a container to grow seedlings in it needs to drain from the bottom.

CONTAINERS

Large pots and containers can be too heavy to move around if you fill them full of earth. Put some old polystyrene in the bottom to keep the container light and also help drainage.

CATS

Cats aren't always unwelcome visitors to your garden. They might love to dig a bit and sometimes roll on your seedlings but on the whole they are a real bonus because they keep mice and birds from stealing your crops.

WATERING

Garden watering systems are cheap and a really excellent idea since they allow you to focus the watering on individual plants. Plus, you can go on holiday and know that your poor garden isn't going to die whilst you are away. The focused watering means that you are not watering the weeds.

Organize some water storage in your garden. Water butts set up to catch the rain water off your roof are an excellent idea – not only do they save money and the environment but they also encourage wildlife into your garden. Our water butts are full of lovely frogs that eat all sorts of undesirable biting insect larvae.

EQUIPMENT

ESSENTIALS

Gardening should be a fun activity, open to all. It's not a sport that requires lots of expensive hi-tech equipment, but you will need some bits and pieces to get going with.

• GARDEN FORK AND SPADE: Get a good quality, lightweight design that is the correct size for you. As a bit of a shortie myself, I get backache almost instantly when I use a normal length fork for digging, so it pays to spend a little time testing and getting the right one for you.

• WEEDING IMPLEMENT: There are lots of different types, from short ones with a metal hook to claw-like implements that double as mini rakes. You'll come to know what style is your favourite.

• HOSEPIPE: Get a heavy duty one and make sure it's long enough to reach to everywhere it needs to go.

• WATERING CAN: A big one with a removable rose on the spout.

• GARDENING GLOVES: Lightweight enough for you to be able to work in but thick enough to protect you from prickles.

• BUCKETS: Sturdy rubberized buckets, with handles, for various jobs including weeding, watering, muckspreading, etc.

• GARDENING SHOES: Yes, it sounds daft but you really do need some sturdy footwear because it's dangerous work for toes when their owner starts forking over hard ground.

At Casa del Sole I inherited a lot of ancient tools, the like of which I have never seen before. It turns out they are antiques, over fifty years old, hand-made and evidently built to last.

• SECATEURS: You'll need these for pruning. I lose mine regularly and have to resort to the kitchen scissors, which are most unsatisfactory. Plus it's important to cut cleanly when pruning rather than hacking with a knife or blunt scissors.

• MUSCLES: If you don't have them already you will either develop them or will have to hire them in! There are some jobs in the garden that require hard work (digging and weeding to name but two). Even labour-saving devices (such as a rotovator) require a good deal of strength to heft them about. On the positive side, your gardening should keep you fit and healthy and all this fresh air certainly helps you sleep at night.

NON-ESSENTIALS (WHICH YOU MAY NEED EVENTUALLY)

• WHEELBARROW: For carting around the large amount of manure you will need.

• WATER BUTTS: Essential for when you start to conserve rainwater and make your own fertilizer.

• IRRIGATION HOSE: After a few weeks of summer watering you'll probably see the sense in investing in an automatic irrigation system.

• GREENHOUSE: Choose one to suit your size of garden – the smallest form of greenhouse being a plastic bag tied around a seed tray or a plastic bottle upturned over a seedling, to an all singing and dancing automatically ventilated and sprinkling glasshouse. I personally have a small, home-made, wooden-framed polytunnel, which isn't beautiful but does the job and didn't cost the earth since all the materials were recycled.

• COLD FRAME: When you have really caught the gardening bug you'll want to extend your season and grow some exotic stuff (like cucumbers), so you will eventually graduate to needing a cold frame. You can buy these ready constructed but the best types are built into the ground.

SOWING SEEDS

Most warm climate-loving plants need as much of a head start as you can possibly give them. In Italy most main crop seeds are planted straight into the garden and thinned out *in situ* because there is no rush since the summer is so long. The easiest way to get a plant going quickly (so it is ready to go outside as soon as the frosts are over) is to sow it indoors in a seed tray. You might have a dedicated greenhouse (with or without heating) or a polytunnel or you might just have a sunny windowsill or porch. It really doesn't matter as long as it is warm (but not near a radiator) and has plenty of natural light. As a rule of thumb, you should aim to find a spot in full sun that is not much more than five degrees warmer than outside but which will not get too cold at night. A heated propagator is useful, especially in unheated greenhouses where the temperature drops sharply at night.

The best thing about sowing seeds in containers, however, is that you can raise them in safety, away from slugs, chickens, cats and any other garden pest that you might have. For this reason, even here in Italy, I sow all my seeds in pots with the exception of carrots, which cannot be transplanted and so have to grow in their bed.

Febbraietto, corto e maledetto

Little February, short and cursed – but it's a good time to plant your seeds inside!

SOWING

You can sow seeds in seed trays or individually in cells. I sow my seedlings in all sorts of recycled containers: yoghurt pots, plastic trays that once held biscuits… the possibilities are endless. You just need to remember to pierce holes in the bottom so there is free drainage. Cells are good – this is a tray of individual little cells designed to take a seed each (though you should plant two in each and remove one if both do germinate). My homespun equivalent is old egg cartons – they work brilliantly.

You can buy special seed compost but I just use the normal stuff and crumble it finely with my fingers. Try to remove all the big lumps so that the seeds have an even compost to grow in.

Most importantly, read the instructions on the packet. These should tell you at what depth, temperature and spacing to sow the seed. You can save yourself a lot of time and effort by following these instructions.

• Seeds should always be sown sparsely. Although you can thin seedlings easily, this also disrupts the roots of the other surrounding seedlings and

In mid-autumn in rural Italy you'll see myriad little fires on the hillside. The olive farmers are tidying their groves in readiness for the olive harvest in late autumn. Mostly what is burning is the spikey brambles – the bane of our lives. We wait until October to do this 'cleaning', as the Italians call it, because any earlier and the dry weather would make setting fires dangerous. In many parts of Italy October is a 'growing month' and you have to leave it until the last minute to cut back the weeds and brambles, otherwise the wretched things grow again and get in the way of your olive nets.

seedlings that have had to compete early on in life often are a little less robust than those given plenty of room (think of them as children).

• Gently firm the compost into your tray or whatever you are using and then make a hole with your finger or a pencil, or a 'pricking out stick' if you have one, to the correct depth and drop your seed in. Smooth the compost over the top of the seeds and crumble on some more if additional depth is needed.

• Make sure you label your seeds or you will end up playing 'guess the seedling' when they start to sprout.

• The seeds now need water. Either mist gently but thoroughly with a spray or place in a tray of water so that the seed containers absorb the water from the bottom. Do not water with a watering can or any other type of pourer; the seed compost is far too fine and shallow and you will displace the seeds.

• For normal watering the easiest thing to do is to invest in some capillary matting, which you saturate with water and stand your seed trays on. The water is absorbed from the bottom upwards, which avoids any seed disruption. You may need to mist the tops of the planters if they look to be drying out. Plus make sure that the matting is always soaked.

GROWING ON

Once the seed has started growing it will put up two 'seed leaves' to begin with. These are deceptive because they normally don't look anything like the leaves that will follow.

• If you have lots of seedlings jostling for space, thin them out (as a rule of thumb a seedling needs at least 3cmsq (1sqin) of soil to itself).

• Wait until you have at least two true leaves before you think about transplanting the seed.

• Once you have a few true leaves you can decide what to do with your seedling. You can either plant it out in the garden if the weather is warm enough or you can pop it in a larger pot to grow on in the warmth for a bit longer. I tend to move mine from the windowsill to the cold frame and then finally into the garden.

• Handling seedlings is a tricky business. They are really fragile and it is easy to crush new leaves or rip roots. If you have less than delicate fingers use a pencil or slim dibber to gently prise away from each other two neighbouring seedlings. Try to hold small seedlings by their seed leaves which, are not as sensitive.

PLANTING OUT

If you plant the seedlings outside, it is a really good idea to cover each little plant with a cloche (I use plastic bottles chopped in half with their tops removed) to protect them from pests and the cold.

Alternatively, if you have a long row of seedlings, a good investment is a mini polytunnel. However, do water well because newly transplanted seedlings have not established their root system yet and need lots of moisture.

Also, when planting outside it is necessary to acclimatize the plant as much as possible, so leave them outside in their seed trays for a day or two before planting them into the colder bed. Plants are enormously susceptible to shock if their climate changes suddenly, so be gentle with them.

After moving seedlings or young plants always water in well.

THE Vegetables

tomatoes

pomodori

What could be more Italian than sun-ripened tomatoes, sweet and still warm from their ripening in the garden. A drizzle of the best olive oil and a bit of ripped basil… heaven. Tomatoes are truly a wonderful crop. The plants produce plenty of fruit and they store well (in the form of chutney, passata and sauce). Home-grown tomatoes are just about as far removed from supermarket-bought ones as you can imagine. Perfect for pots, patios, windowsills, or grow a field full… you'll never have enough.

There are the bush varieties, known as 'determinate', which are low-growing and don't need staking, or the 'indeterminate' type, which need to be secured to a cane or trellis. On the seed packet or seedling tray it will say what variety you have chosen.

SOIL/SITE

Full sun, richly fertilized soil with plenty of calcium. They grow adequately in a grow-bag but you can provide them with deeper rooting and better irrigation in a decent-sized trough full of well-rotted manure. You can start tomatoes inside on a window ledge and in a greenhouse and they will also do well outside, preferably on a south-facing wall. Some varieties of cherry tomato grow in cascades and are stunning in hanging baskets (but they are a nightmare to keep well watered in summer – you really need an automatic feeder).

SOWING

Depending on the variety, seeds should be sown in early spring – check the packet as some require an earlier sowing. The final planting distance is about 30cm (12in) apart. Whatever you do don't put them out until you are sure there isn't going to be any more frost locally. Harden the seedlings off first (putting them outside during the day and bringing in at night for a week) before you plant them… they can get very upset at a sudden change in temperature.

TREATMENT

Keep weed-free until the plants reach a height where they will not have to compete with the weeds for light. Mulch plants to help retain moisture. Don't let the soil dry out but also make sure they are well-drained (fussy beggars!). This is where a drip water feeder comes into its own.

'Indeterminate' varieties will need all the side shoots pinching out on a regular basis. Side shoots appear at the junction of a leaf and the stalk – all you have to do is nip it off with your fingernails when it's still small. If you leave them the plant becomes too heavy and leafy with all the nutrients going towards growing new plant rather than new fruit. You won't have a slug problem with tomatoes but keep the chickens out of your garden once they start to ripen, as the naughty chooks delight in eating them straight off the vine.

HARVESTING

Pick as soon as the tomatoes are ripe. If at the end of the season you are left with green tomatoes on a vine, remove the vine with the tomatoes attached and place on a sunny windowsill – they might just ripen but if not then you can use them in chutney.

STORAGE

Fresh tomatoes don't store for very long. However, they are perfect for bottling and making into chutney (see p.163).

EXPECTED YIELD

Yield is heavily dependant on variety and conditions but expect 2–3kg (4½–6½lb) per plant.

VARIETIES

These are the best from hundreds of available varieties:

For meaty tomatoes try our local 'Costoluto Fiorentino'. These plants produce large misshapen fruits that taste sublime. They should be grown in the greenhouse or outside only in the warmest, sunniest of locations.

My favourite plum tomato is 'Roma', which is one of the hardiest of the Italian varieties. You should be able to grow it outside and the fruits are good for eating fresh or making passata (see p.167).

Cherry tomatoes are the little ones that are so sweet you can't stop eating them. 'Lilliput' is one of my favourites because it produces absolutely loads of fruit and is easy to grow inside or out. You can grow it in hanging baskets and let the vines drape downwards, or in pots or in the ground and treat like a normal determinate variety. Cherry tomatoes are the easiest of all the tomatoes to grow and so if you only have room for a couple of plants make sure one of them is a cherry.

SERVES FOUR

6 ripe tomatoes (it doesn't matter
    what type – just make sure they
    are luscious)
4 balls of buffalo mozzarella
bunch of basil
top-quality olive oil
salt and freshly ground black pepper
top-quality balsamic vinegar

caprese salad
insalata caprese

This quintessential Italian summer dish is only worth eating when prepared with the perfectly ripe tomatoes of summer, fresh basil straight from the bush and good-quality buffalo mozzarella.

Slice the tomatoes thickly and do the same with the cheese. Arrange on a large serving dish as artistically as you like. Rip the basil leaves (do not chop – ripping is better for releasing the flavour and scent) and sprinkle them on top of the tomatoes and cheese.

Drizzle a little olive oil over the salad and season with a bit of salt and fresh ground black pepper.

Leave the olive oil and balsamic on the table so that your guests can decide how much dressing they like.

tomatoes stuffed with rice

pomodori ripienie

This recipe is brilliant for using up leftover rice. The herb mix in the ingredients is what I like – you could choose chives, parsley or whatever you have in the garden. So simple but tasty and filling.

If you use pre-cooked rice, the result will be more moist but be really careful when removing the tomatoes from the dish since they are more likely to collapse.

SERVES FOUR

4 large ripe tomatoes – but not wrinkly
    and over-ripe or they will collapse
50g (2oz) long grain rice or leftover
    pre-cooked rice – enough to
    three-quarters stuff the tomatoes
1 tbsp chopped fresh basil
1 tbsp chopped fresh mint
1 tbsp chopped fresh oregano
2 tbsp olive oil
rock salt and freshly ground
    black pepper

Preheat the oven to 200°C (400°F).

Slice the tops off the tomatoes with a very sharp knife and put on one side, as you will be replacing this little lid.

Using a teaspoon, scoop out the seeds gently and put on one side in a small bowl. To this bowl add the remainder of the ingredients and stir to combine.

Spoon this mixture into the tomatoes, replace the lids and place in an ovenproof dish. Cook for about 45 minutes, basting a few times with the juice in the dish.

balsamic cherry tomatoes

pomodorini al balsamico

As you might have guessed, summer is a time of lots of tomatoes at Casa del Sole. It is important to use recipes that are very different in order that we don't get tired of a particular crop. These roast tomatoes with that wonderful balsamic flavour are a delight. This is a tasty side dish – serve with crusty bread for dipping.

PER PERSON

1 'bunch' of cherry tomatoes on
    the vine
drizzle of oil
drizzle of balsamic vinegar
pinch of salt

Preheat the oven to 225°C (425°F). Place the tomatoes, still on their vines if possible, in a baking tray. Drizzle the oil and vinegar over the top of the tomatoes – enough to make a very shallow puddle in the bottom of the dish. Sprinkle with salt.

Roast in the oven for about 20 minutes or until cooked through. Serve with the oil and balsamic mixture from the bottom of the dish.

bread and tomato salad

panzanella

Many Italian recipes are 'use up the leftovers' based. This lovely salad traditionally uses up the remainder of yesterday's bread. The meaty 'Costoluto Fiorentino' variety of tomato are perfect for this salad (see p.52).

SERVES SIX

500g (1lb 2oz) ripe tomatoes

1 small red onion, peeled and sliced

2 celery stalks, diced

slurp of red wine vinegar

couple of glugs of olive oil

salt and freshly ground black pepper

500g (1lb 2oz) dry white bread

big bunch of basil and sprig of
    fresh herbs

Cut the tomatoes into small chunks and put in a bowl with the onion and celery. Sprinkle the vinegar over. Add the olive oil and season well with salt and pepper.

Roughly crumble, cube or slice the bread, then put in a bowl and pour enough water over to cover it. Leave it for less than a minute and then drain into a colander. Squeeze out all the excess water. Add the bread to the tomatoes. Rip up the basil leaves and stir well. Garnish with fresh garden herbs of your choice.

'bolognese' sauce

ragu

Ragu is what foreigners might term 'Bolognese' sauce. It is a signature Italian sauce used for pasta, gnocchi, lasagne and polenta. A good ragu takes a while to make (although the process is very simple), so I make a large batch and freeze it in portions for quick dinners and easy-make lasagne.

SERVES SIX IF USED
WITH SPAGHETTI

large slurp of olive oil

2 large onions – red or white, peeled
    and diced

2 large garlic cloves, peeled and minced

4 celery stalks, diced

175g (6oz) pancetta or streaky bacon
    (smoked is fine), diced

500g (1lb 2oz) minced beef

1 tbsp plain flour

425g can of chopped tomatoes (if you
    have fresh tomatoes to use up then
    use 500g/1lb 2oz of them and dice
    them – the taste is excellent when
    the ingredients are all fresh)

250ml (9fl oz) water or red wine

1 beef stock cube

1 tbsp fresh oregano or 2 tsp dried

½ tsp grated nutmeg

salt and freshly ground black pepper

Slurp a big glug of olive oil into a large, heavy-based pan and heat. Soften the onions, garlic and celery in the pan for a minute or so. Add the pancetta or bacon and fry for a further couple of minutes, then add the mince. Once the meat has browned, sprinkle on the flour and stir in. Tip in the tomatoes and the water or wine and crumble in the stock cube. Stir to combine and then add the oregano and nutmeg.

Cover the pan and simmer on a very low heat for 45 minutes, stirring occasionally to stop it sticking to the bottom and burning. After 45 minutes taste the sauce and season with salt and pepper as you deem fit.

Serve with gnocchi, pasta quills or polenta… not with spaghetti if you want an authentic Italian dish.

tomatoes on toast

bruschetta al pomodoro

This is such a quick and easy dish to prepare that I often serve it when people pop around for a glass of wine and a chat. It's also good for when you have a large dinner party, as people can nibble on the bruschetta whilst you get on with the cooking; it's a traditional Italian antipasto.

PER PERSON

1 medium-sized tomato – the fleshy
    variety, diced

1 red onion, peeled and diced

bunch of basil

small slurp of excellent olive oil

few drops of aged balsamic vinegar

1 pinch of salt

2-day old crusty bread, sliced thickly

Combine the tomato and onion in a bowl. Set aside a sprig of basil for the garnish and rip the remainder into shreds. Add the basil, olive oil, vinegar and salt to the bowl, mix well and leave in the fridge for a couple of hours for the flavours to infuse and the water to come out of the tomatoes.

Toast the bread just before you are ready to serve. Spoon the tomato mixture on top and garnish with the sprig of basil.

leaves & salads

lettuce

Often viewed as the base of any salad, lettuce comes in many different varieties. In Italy it is grown for winter consumption because it likes cool and moist conditions but in cooler climes it is best sown for a summer/autumn harvest. Or you can sow in autumn as long as you have a heated greenhouse.

SOIL/SITE
Any soil that is well-drained – lettuce loves manure!

SOWING
Sow in drills about 2cm (¾in) deep from mid-spring (after threat of frost) to late summer. Thin to 30cm (12in) apart.

TREATMENT
Cover the later sowings with a polytunnel and get a crop in autumn.

AFTERCARE
Watch out for slugs, the bane of the leaf grower's life.

HARVESTING
Simply pick off the leaves as you need them. Be careful not to over-pick a plant – they need some leaves left to grow more.

STORAGE
Lettuce needs to be picked and eaten immediately.

EXPECTED YIELD
1kg (2¼lb) per msq (sq.yd).

chicory

cicoria

One of the easiest winter salads to grow, and crunchy and tasty too. The bitter taste of chicory goes perfectly with seafood and complements any salad.

SOIL/SITE

Any soil that is well drained.

SOWING

Sow in drills about 2cm (¾in) deep from early to late summer. Thin to 30cm (12in) apart.

TREATMENT

In late autumn, cut the plant to just below the crown and, if you want to have a continuous supply through winter, replant in pots and keep in a frost-proof shed or cupboard. They will then shoot again.

HARVESTING

Break the shoots off as you need them – they will keep growing every month.

STORAGE

Once picked you can't store it, so only harvest when you want to eat.

EXPECTED YIELD

1kg (2¼lb) per msq (sq.yd).

VARIETIES

In a cooler climate I would grow 'Radicchio Rossa di Treviso Tardiva'. This variety is really quite resistant to frost and so can be harvested right the way through winter. 'Radicchio Grumolo Rossa' is another good variety that resists frost well and with this one you cut the head once and leave the plant in the ground and it will grow again. Good value!

rocket

rucola

This is a super tasty addition to any salad. Rocket is easy to grow, and easy to prepare for the kitchen – it just needs a wash.

SOIL/SITE

Any soil that is well drained.

SOWING

Sow in drills about 1cm (½in) deep from mid-spring (after threat of frost) to late summer. Thin to 20cm (8in) apart.

TREATMENT

Cover the later sowings with a polytunnel and get a crop well into winter.

AFTERCARE

Watch out for weeds and slugs.

HARVESTING

Simply pick off leaves sparingly as you need them. Be careful not to over-pick a plant – they need some leaves left to grow more.

STORAGE

Rocket won't store well even in the freezer so look at it as a seasonal treat.

EXPECTED YIELD

1kg (2¼lb) per msq (sq.yd).

VARIETIES

There are two types of rocket – cultivated and wild. Cultivated needs to be re-sown after you have harvested the crop. With the wild variety, you can leave a few plants in the ground and it will re-seed itself. It's also much more peppery than the cultivated variety. I grow this type in its own dedicated bed and simply keep it weed-free for a wonderful low-maintenance crop.

spinach

spinaci

I viewed spinach as a northern plant because it seems to like such a lot of water but Italians love this versatile plant. Eaten raw in salads, used as a stuffing for pasta, put on pizza or (my personal favourite) simply boiled, drained and served with garlic, chilli oil and lemon juice.

If you aren't a spinach fan then it's probably because you've never had home grown and properly prepared spinach. Not only is it easy to grow and will provide a crop most of the year, but it tastes great too!

Spinach likes fertile soil, plenty of water and partial shade. It is perfect for intercropping with tall plants such as sweetcorn or sunflowers since it likes the shade. You need to plant quite a lot of it for a good few meals but it is a 'cut and come again' crop and so provides a good batch a couple of times over on the same stalk. There are enough seasonal varieties for you to grow spinach all year around. It likes plenty of room and so isn't a great container garden plant.

SOIL/SITE

Richly manured soil, not too sunny.

SOWING

Depending on the variety sow when advised on the packet. Outdoors: sow in drills 2.5cm (¾in) deep and 30cm (12in) apart. Thin when seedlings are about 2cm (¾in) high to about 15cm (6in) between each plant. In trays: Sow a few seeds per cell and thin out to one robust plant when about 2cm (¾in) high.

TREATMENT

Keep weed-free by hoeing, water in summer so that the soil is never dried out. Mulch in summer to help retain moisture. Watch out for slugs and snails.

RIGHT: The kitchen garden in early spring is normally in slight disarray unless you are punctilious. Here you can see some of my raised beds in dire need of some digging and weeding.

HARVESTING

Pick the leaves whilst they are young and tasty. Pinch out the sprouting heads and eat them – they should grow back double. Or you can chop the whole plant at about 3cm (1in) from the ground when it is quite big, use those leaves and then wait for the stump to sprout again. Don't forget to keep it weeded and watered.

STORAGE

Intended to be eaten fresh, but the best way to preserve spinach is by wilting it in boiling water, compacting it into freezer bags and freezing for future use.

EXPECTED YIELD

1kg (2¼lb) per msq (sq.yd) per growth cycle (i.e. if you cut it and wait for it to grow back you get double this amount from your area).

VARIETIES

'Riccio d'Asti' is a variety that grows well in cooler climates and doesn't take up too much room.

radishes

ravanelli

In gardening, instant gratification is pretty unusual. However, radishes are as near as you can get to this and thus are brilliant for children to grow. The seeds sprout within a few days and, depending on conditions, you can be picking them within a month of planting.

SOIL/SITE

Any soil at all… if you can't grow radishes then hang up your trowel and find another hobby!

SOWING

Sow in drills about 1cm (½in) deep from early spring (and hope there aren't any hard frosts) to late summer. Thin to 10cm (4in) apart. Carry on sowing through summer for a constant crop.

TREATMENT

Keep weed-free and watch out for slugs. Water well or they taste a bit woody.

AFTERCARE

Don't let them bolt (run to seed) – pick as soon as they are ripe.

HARVESTING

Simply pick them out of the ground, give them a wipe and eat them – delicious.

LEFT: Radishes are a joy to grow since they are so speedy to mature. Even the most impatient gardener will be amazed at how quickly they are eating the fruits of their labour.

STORAGE

They don't store… pick 'em and eat 'em.

EXPECTED YIELD

1kg (2¼lb) per msq (sq.yd).

VARIETY

'Rapid Red Sanova' – looks and tastes just like a radish should!

steak with rocket

tagliata con rucola

Rocket is a superbly spicy leaf that greatly complements this simple steak. Italians like their steak cooked on the outside and bloody within. To eat it 'ben cotto' – well cooked – is most unusual and often you will find that an Italian chef resists your request for a well-done steak. Behind closed doors it's another matter…

SERVES TWO

2 very thick sirloin steaks

balsamic vinegar or balsamic glaze

sea salt and freshly ground black pepper

2 big handfuls of rocket

olive oil

Heat the griddle on a high temperature and place the steaks to cook for 5–7 minutes on each side for a rare steak. Sprinkle with vinegar on each side towards the end of the cooking process.

Sprinkle a chopping board with sea salt and black pepper, then put the steaks on the board and cut them into strips.

Put a mound of rocket on each plate, arrange the meat on top and then top with more pepper. Serve with access to oil and balsamic vinegar or glaze.

green salad

insalata verde

A lovely crisp green salad is often served with a *secondi* (main dish) in a side dish. The salad is never placed on the plate with the main dish but eaten separately from the bowl. Italians like to simply drizzle a little good olive oil and red wine vinegar on their green leaf salad. Balsamic is more the reserve of the tomato, and complicated dressings are hardly ever found in traditional Italian cookery.

PER PERSON

big bunch of salad leaves (including rocket if you have some – for a lovely Italian kick)

red wine vinegar

extra virgin olive oil

Clean the leaves and dry. My friend Antonella showed me how to dry leaves without bruising them by placing them in a dry tea towel, gathering the corners and gently swirling it over your head.

Place in a bowl and serve. Let your guests add the vinegar and oil to their own taste.

spinach and ricotta cannelloni

spinaci e ricotta cannelloni al forno

These enormous tubes of pasta are perfect for stuffing with a thick sauce. Because they are cooked in the oven they are easy to prepare as *primi piatti* (starters) because you can do everything in advance and pop them in the oven at the last minute. The word ending 'oni' indicates something is big. If the word ending was 'ini' it would mean small.

SERVES FOUR

8 cannelloni tubes

30g (1¼oz) freshly grated Parmesa
    for topping

FOR THE STUFFING:

600g (1lb 5oz) clean spinach
    leaves, blanched in boiling water
    and drained

250g (9oz) very fresh ricotta

½ tsp freshly grated nutmeg

½ tsp salt

splash of olive oil

4 tbsp cream

2 eggs, beaten

grind of black pepper

FOR THE BECHAMEL
    SAUCE:

40g (1½oz) butter

40g (1½oz) plain flour

500ml (18fl oz) milk

2 bay leaves

¼ tsp freshly grated nutmeg

30g (1¼oz) freshly grated Parmesan

Preheat the oven to 175°C (350°F). If using dried cannelloni, bring a large pan of water to the boil and drop in the tubes. Cook for about 5 minutes, then check that they are on the hard side of 'al dente' – be very careful not to overcook them or they will be very difficult to stuff and will rip easily. You will know when they are done because you can bend them a bit but they aren't flaccid.

Meanwhile, combine all the stuffing ingredients in a large bowl and mix thoroughly.

Once the pasta is cooked, remove from the water and drain. Leave to cool until you can handle it. Stuff the tubes with the ricotta and spinach mixture and put in a large ovenproof dish.

Now make the sauce. Melt the butter in a heavy-based pan and then add the flour, beating well until it is a thick paste. Add the milk a drop at a time and stir/beat constantly. As soon as the mixture becomes a liquid add the rest of the ingredients and carry on adding the milk. Once all the ingredients are combined keep cooking on a low heat and stirring (lumpy sauce is usually because you don't stir enough – don't put the spoon down for a moment if you want a lovely smooth sauce!) for another 5 minutes.

When the consistency of the sauce is correct (thick and creamy) take off the heat, fish out the bay leaves and pour over the cannelloni. Top with the remaining grated Parmesan and cook in the oven for about 35 minutes until the top is golden brown.

RIGHT: This creamy dish turns simple spinach into something really special. Spinach is a late-autumn and winter crop and is full of vitamins and minerals.

courgettes & squashes

butternut squash

zucca

Butternut squash is my favourite type of squash because it tastes so wonderful in soups and risotto. The slightly sweet nutty flavour makes it a hit with the children too. And the best news is that it's easy to grow and should give an excellent ratio of food to effort.

The following method of growing is the same for any pumpkin or squash.

SOIL/SITE

They grow pretty much anywhere but they prefer to be planted in plenty of muck. They'll grow on your muckheap if you let them – and why not? Otherwise, any soil that has been heavily manured.

SOWING

Sow inside in trays from about late winter and plant out under cloches or upturned plastic bottles or jam jars when the first couple of true leaves show (these are the ones that look star-shaped rather than the rounded baby leaves). They will withstand a light ground frost if they are covered but it's better to keep them inside until the risk is gone. You can sow them *in situ* outside from late spring onwards for a later crop.

TREATMENT

Keep weed-free to begin with. This type of squash is a wonderful climber, so set it off over your shed roof but be prepared to have to support the fruit once it gets big. It's just as happy meandering along the ground, however, but needs a lot of room.

AFTERCARE

Watch out for slugs eating the fruit. Keep the squashes off the ground by putting a flat stone, tile or somesuch underneath them. This prevents any rotting of the flesh on wet ground.

HARVESTING

When the leafy bits of the plant start to die is the time to pick the squash. If there is a risk of an early frost then cut your losses and bring in the fruit early – frost ruins ripe squash.

STORAGE

Butternut squash store brilliantly right the way through winter. Simply keep them hung – or put them on a rack – in a dark, cool and airy place.

EXPECTED YIELD

2–3 squashes per plant.

courgette
zucchini

Courgettes are marrows that have not been allowed to grow big. In my first year of gardening I planted six plants and they were satisfyingly prolific, but we were all so sick of courgette that year that I didn't plant any the following spring. Now, I put in three plants and that is perfect for us. They are very easy to grow.

SOIL/SITE
Sheltered site in full sun, well-fertilized, well-drained but continually moist soil.

SOWING
Sow in seedbeds or pots throughout spring (as a rule but varieties vary so follow the instructions on the packet). Harden off well if planting outside. Plant out about 70cm (28in) apart when the seedlings have five or six good leaves. Protect the young plant with a jam jar or plastic bottle – mostly from slugs and snails. Can be planted in a wide container but don't put it anywhere that you have to brush past it, as the leaves and stems are spikey.

TREATMENT
Keep weed-free until the plants get to a height that they smother the weeds themselves. Mulch in summer to help retain moisture. Watch out for slugs and snails on young plants, later the prickles will keep most predators at bay.

AFTERCARE
Keep them moist but not wet. If growing marrows, once the fruit is large and lying on the ground, protect from rot by placing a tile under the fruit. If you are not eating the flowers, once they have died off remove them since in wet climates the flower rotting can spread to the tip of the courgette and spoil it.

HARVESTING
Harvest tiny with the flowers on for a deep-fried Italian feast, later when they are tender, or leave on the plant to grow into marrows for stuffing. Use a sharp knife to chop through the stems. The more you pick, the more come along.

STORAGE
You can freeze them, chopped, but they go mushy when defrosted, so are only good for stews or ratatouille. Marrows tend to keep for a month or so if hung up in a cool, dark environment.

EXPECTED YIELD
15+ courgettes per plant.

deep-fried courgette flowers stuffed with ricotta and chilli peppers

fiori di zucca

This recipe was kindly given to me by a chef who helped me learn a lot about Italian cookery. He worked at the time in a large villa outside Lucca in Tuscany and this was the first recipe we cooked together. I had never deep-fried anything in my life before, so it was a revelation how easy it could be. The batter is a little bit more complicated than a normal batter but it comes out so amazingly crispy that it's worth taking the extra few minutes over.

SERVES FOUR

12 very fresh, tightly closed
    courgette flowers

oil for deep-frying

FOR THE BATTER:

50g (2oz) plain flour

¼ tsp salt

2 tsp olive oil

75–100ml (2½–3½fl oz) cold
    fizzy water

3 egg whites

FOR THE FILLING:

100g (3½oz) ricotta

pinch of salt

1 hot chilli, deseeded and
    finely diced

First make the batter: sift the flour and salt into a bowl. Add the oil and mix into a paste. Then add the water slowly, changing to a whisk when the mixture becomes liquid enough. In a separate clean bowl, whisk the egg whites (use a clean whisk too or they will never whisk up properly) until they are stiff and glossy. Then fold the egg whites into the batter. Place in the refrigerator until you are ready to use it.

For the filling, put the ricotta, salt and chilli into a bowl and mix in well.

Take the courgette flowers and and trim the stalks to about 3cm (1in) in length (so that you have something to hold onto). Then carefully spoon the filling into the flowers. Don't overfill them or the stuffing will come out when you are cooking.

Pour sufficient oil into a deep pan or fat fryer deep enough to cover the top of the flowers and heat to about 180°C (350°F) – you can test whether it's ready by dropping a drip of batter in: if it bubbles and cooks quickly then the oil is ready. Dip a courgette head in the batter, coat fully apart from the stem (which you don't normally eat but use to pick them up with) and place in the oil immediately. Don't crowd them in the oil or they won't cook well. They will be done in a matter of minutes and are ready when the batter is golden brown. Fish out with a slotted spoon and place on kitchen paper to drain. Serve immediately to loud applause.

spaghetti with baby courgettes
spaghetti ai zucchini

Zucchini (courgette) are astoundingly gorgeous when eaten tiny and tender. In their season you normally have so many of them that it is no shame to eat them small.

SERVES FOUR

1 tsp salt

500g (1lb 2oz) spaghetti

plenty of olive oil

about 12 baby courgettes, thinly sliced

2 garlic cloves, peeled and diced

freshly ground black pepper

bunch of basil

freshly grated Parmesan to serve

Bring a large pan of water to the boil for the pasta and add the salt. When the water is boiling pop the pasta in to cook.

Meanwhile, heat a glug of olive oil in a frying pan. Gently cook the courgettes and garlic in the oil for a few minutes.

Drain the pasta and add to the courgettes. Season with black pepper and toss to coat all the pasta in oil. Add more oil if the pasta seems dry. Roughly rip the basil leaves and toss them into the pasta mix. Serve with plenty of Parmesan on hand.

pumpkin and sage ravioli
ravioli di zucca e salvia

This is a wonderful winter dish packed with vitamins and full of flavour from your garden – even in the middle of winter. The fresh pasta is wonderful, not tricky – just a little messy. You can make a big batch and freeze some for another day. Pasta should really be made on a flat marble worksurface, but this is not vital.

SERVES SIX

FOR THE PASTA:
500g (1lb 2oz) plain flour
4 eggs
chilled water

FOR THE FILLING:
1kg (2¼lb) butternut squash, cubed
slurp of olive oil
salt and freshly ground black pepper
100g (3½oz) ricotta
30g (1¼oz) freshly grated Parmesan
1 tsp freshly grated nutmeg
fistful of sage, stalks removed
    and chopped
1 egg yolk, beaten

FOR THE SAUCE:
250g (9oz) butter
20 sage leaves

Preheat the oven to 180°C (350°F). Put the butternut squash in a roasting tin and coat with a good glug of oil. Season with salt and pepper and roast for about 40 minutes.

Meanwhile, make the pasta. Mound the flour on your worksurface and make a well in the centre. Break in the eggs and whisk with a fork, incorporating the flour around the edges as you go. The dough should be soft, pliable and dry to the touch. Add some chilled water if the mix is too dry. Knead for 5–10 minutes until the dough becomes glossy and elastic. Cover with a tea towel and leave for 30 minutes.

Remove the squash from the oven and set on one side to cool. When cool enough to handle, add the ricotta and mix in all the remaining filling ingredients except the egg yolk.

Cut the pasta into quarters. Using a wooden rolling pin, roll out one-quarter to about 5mm (¼in) depth and then fold it over once and repeat this for a traditional 10 times. Then finally roll the pasta out to the thickness you require, about 2cm (¾in). Repeat this with the next quarter. Try to get the finished sheets the same size and shape. Take the first sheet of pasta and spoon 1 teaspoon blobs of filling, 3cm (1in) apart, in a straight line down the sheet. Repeat this so you have a grid of filling blobs. With a pastry brush, brush the egg along all the avenues of the grid. Place the second sheet of pasta on top of the first and settle gently over the stuffing. With your fingertips, gently press the top layer of pasta down to the bottom layer along the avenues where there is no stuffing. Then, using a sharp knife, cut the ravioli along the empty avenues (between the blobs of filling) in one direction and then in the other to make squares. Repeat for the second two batches of pasta. The pasta can be cooked immediately or stored in an airtight container in the fridge.

Bring a large pan of salted water to the boil and, once it is boiling steadily, drop the ravioli into the water. The pasta should take only 3–4 minutes to cook and will rise to the surface when it is ready.

To make the sauce, heat a little of the butter in a pan, being careful not to burn it. Add the sage leaves and fry until crisp. Remove and then melt the rest of the butter in the pan. Add the sage leaves to the butter and pour sparingly over the ravioli.

sweet & chilli peppers

Peppers are a must for the Italian garden. My friend Marisa grills them and preserves them in oil. But I never seem to grow enough for this and we eat them, usually still warm from the sun, in salads or baked and stuffed.

Peppers are really a warm climate plant but will grow well in cooler climates if you coddle them a bit. They are best grown in a greenhouse, conservatory or windowsill because they like it warm, resent too much rain and love plenty of sunshine. Don't grow sweet peppers and chilli peppers near each other – if they cross-pollinate the sweet peppers will take on a spicy tang!

SOIL/SITE

Sunniest, most sheltered site with moist, well-drained soil, preferring a slightly acid environment.

SOWING

Early to mid-spring. It really is best to sow in pots inside. Plant out sweet peppers about 40cm (16in) apart or in a pot not less than 40cm (16in) diameter when the plants are about 20cm (8in) high. Chilli peppers tend to be smaller bushes – check the packet and proceed as advised. Harden off carefully. Any frost or sudden drop in temperature will kill off your crop.

AFTERCARE

Balance the fine line between keeping the soil moist but not wet. If under-watered the flesh of the peppers will be thin and hard, but they don't like it too wet. Mulching is good for moisture retention. Watch out for the plant becoming overburdened with fruit – you might need to prop it up with twigs.

HARVESTING

All peppers start off green. If you leave them they will go red or yellow (unless they are a specific green variety). Pick when they look about the right size. The more you pick, the more flowers are produced and so the more will grow.

STORAGE

Chilli peppers dry well and can be stored in this way. I also freeze some ready

chopped, as the frozen ones retain their strength better than dried ones. Sweet peppers can be preserved in oil or chutney but otherwise won't store more than a few days.

YIELD

Very dependant on conditions and variety but around 8 peppers on a sweet pepper plant and 20+ chillies per chilli bush.

VARIETIES

'Corno Rosso' are the easiest to grow and will even do well outside in a sunny year if planted in a sheltered spot. They are a little unshapely, though, so if you want a good pepper for stuffing grow 'Topepo'.

spicy spaghetti with garlic

spaghetti a l'aglio, olio e peperoncino

For a long time I avoided this dish because it sounded too plain. What a mistake! This is Tuscan cookery at its simple best. The best oil, the best garlic, the best chilli. Perfect.

SERVES SIX

500g (1lb 2oz) spaghetti

about 250ml (9fl oz) good-quality extra virgin olive oil

3 dried hot chillies

(if you have it, substitute the two above ingredients for one cup of chilli-infused oil)

3 garlic cloves, peeled and minced

salt

handful of chopped parsley (fresh or from the freezer)

50g (2oz) freshly grated Parmesan plus extra to serve

Cook the spaghetti as normal.

Heat the oil in a small pan. Add the garlic and crumble in the chillies, then fry for a few minutes.

Drain the pasta and return to its pan. Add the garlic and chilli mixture and season with salt. Add the parsley and Parmesan and toss. If the pasta seems dry then add more oil.

Serve with extra grated Parmesan.

hot pepper pizza
pizza peperoni al diavolo

The Italian word for a sweet pepper is *peperoni*. When I first moved to Italy I didn't speak any Italian and kept being upset when I ordered a pepperoni pizza because I was expecting spicy salami. I have no idea how the word has become so corrupted but I do know that the pepperoni pizza is now one of my favourites. Because Italians like their flavours clean and uncomplicated, they don't traditionally put a lot of different toppings on the same pizza. The 'Diavolo' bit means that it's hot as hell!

Pizza in Italy is made in large stone ovens, which gives them a unique taste. However, all you need to do is put your oven up to the hottest setting and you can cook a pretty good pizza 'a casa'.

I always make my pizza dough in a bread-making machine – it's easy and you can get on with other things whilst the machine does the hard work.

MAKES THREE PIZZA

FOR THE DOUGH:

250ml (9fl oz) water

2 tbsp caster sugar

1 tbsp (or sachet) dried yeast

700g (1½lb) 'oo' flour
(or breadmaking flour) plus
extra to dust

1 large slurp of olive oil plus
extra to grease

1 tsp salt

FOR THE PIZZA:

125ml (4fl oz) thick passata

1 tsp salt

1 pepper, seeded and thinly sliced

300g (11oz) mozzarella, cut into
small dice

drizzle of very strong chilli-infused
olive oil

dried or fresh oregano to taste

To make the dough: put the water, sugar, dried yeast, flour, oil and salt in the bread machine and choose 'dough setting'. Do it in this order because the yeast doesn't like to be in direct contact with the salt.

If you don't have a bread machine, follow the instructions above in a bowl and mix in until you have dough. Then knead the dough on a floured surface for about 10 minutes until it becomes soft and elastic. Leave for 10 minutes and then knead for another 10 minutes. Leave in the bowl in a warm place, covered with a teacloth for about 1 hour.

Meanwhile, make the pizza sauce. Combine the passata, salt and oregano and leave at room temperature to infuse.

When the dough is ready, chop it into three equal lumps. At this point you can shape it into little balls and leave it to rise again under a cloth until you are ready to use it, or you can use it immediately. Preheat your oven to its highest setting.

Lightly grease your pizza pan with oil – use any non-stick flat-bottomed baking sheet or pie dish. Roll each pizza dough out to size, then lift onto the pan. Push out using your fingers if the pizza dough is too small.

Spoon 3–4 spoonfuls of the passata mixture into the centre of each pizza and use your spoon in increasing circles to push the sauce out to the edges. Then arrange the pepper slices on top, sprinkle the cheese evenly over and place in the hottest part of your oven. Depending on your oven temperature, they should take between 5–15 minutes to cook. You'll know they're done when the edges of the base start to brown and when shaken the pizza should rattle around in the tin. The mozzarella will be melted too.

Remove from oven and drizzle well with the infused oil and serve immediately.

stuffed peppers
peperoni ripieni

Home-grown peppers are often smaller than the huge shop-bought ones but they are twice as tasty. Peperoni Ripieni can be a *primi piatti* (starter) without the meat or a *secondi* (main course) with the meat, unless you are vegetarian, in which case substitute with chopped walnuts and pop in an extra free-range egg.with oil with oil

SERVES SIX AS A STARTER
OR ANTIPASTI – DOUBLE
FOR A MAIN COURSE

3 large peppers
splash of olive oil
1 small red onion, peeled and
    finely chopped
1 garlic clove, peeled and minced
150g (5oz) breadcrumbs (see variation)
1 egg
30g (1¼oz) freshly grated Parmesan
1 tomato, chopped
2 sprigs of chopped herbs from the
    garden – use whatever you have
salt and freshly ground black pepper
150g (5oz) mozzarella, cubed
extra virgin olive oil to serve

VARIATION
You can replace half the breadcrumbs
with the same volume of sausagemeat
or mince if you prefer a meaty version.

Preheat the oven to 170°C (325°F).

Cut the peppers in half, through the middle of the stalk and remove the seeds. Place on a baking tray.

Heat a pan with a little oil and cook the onion and garlic until translucent. Place in a large bowl and stir in the breadcrumbs, egg, Parmesan, tomato and herbs (and any meat if you are using it). Season with salt and pepper to taste. Fill the pepper halves almost to the top with the stuffing. Place the mozzarella cubes on top and bake for about 40 minutes.

Add a drizzle of extra virgin olive oil and serve.

brassicas

In many parts of Italy, the winter months are very cold. The kitchen garden always has something in it, though. Brassicas (cabbage, cauliflower and broccoli) are a winter staple. There is no reason why you can't grow these things in a cooler climate. Broccoli is frost-resistant, as is cavolo nero, which has kept many a Tuscan family alive during winters in the war. If you live in a place where the ground freezes for weeks on end you can grow brassicas in an unheated greenhouse.

broccoli

I just love broccoli. It's a wonderfully reliable winter plant and is fairly easy to grow. Traditional broccoli is great and I plant plenty but also try sprouting broccoli, which is even more hardy through winter and carries on sprouting when you cut the first crop – perfect for smaller gardens or containers. Apparently, birds can be pests since they also get hungry in winter. We have five good farm cats so don't have any problem at all but if necessary just rig up a net over the bed when the heads mature.

SOIL/SITE
Full sun, richly fertilized soil slightly alkaline (make sure you have added a little lime before planting).

SOWING
Sow in drills or pots from late spring (as soon as the weather starts to warm up). Plant out about 60cm (24in) apart when the seedlings have five or six good leaves. For sprouting broccoli, keep planting up to mid-summer to get a continuous crop.

TREATMENT
Keep weed-free (hoe) until the plants reach a height where they will not have to compete with weeds for light (in Tuscany the weeds stop growing in the autumn). Mulch in summer to help retain moisture. Watch for slugs and snails.

HARVESTING
Depending on what you have planted (autumn, winter or spring varieties), you

can cut from early autumn to late winter. You can tell they are ready because the flowers look like broccoli! Don't leave it too long before cutting the flowers since they can open up and not be as tasty. For sprouting varieties, cut as soon as the sprouts are big enough to be worth eating and keep cutting until the plant doesn't produce any more.

STORAGE

Broccoli doesn't store well in the freezer – it's grown as a winter crop so you shouldn't need to store it anyway.

EXPECTED YIELD

You should get one large head per 'hearting' broccoli and numerous shoots from 'sprouting' varieties.

cabbage

cavolo

Cabbage is used in Tuscan cookery but the varieties are limited. You either get the large Savoy cabbage, which is easy to grow but takes up an awful lot of space, or more popular is the 'Tuscan' cabbage, which is really more like kale than cabbage but tastes wonderful. It grows tall with long narrow leaves so doesn't take up much room.

SOIL/SITE

Any soil normally does for this hardy plant.

SOWING

Sow in drills or pots from late spring (as soon as the weather starts to warm up). Plant out about 60cm (24in) apart when the seedlings have five or six good leaves. Keep planting into mid-summer for a winter crop.

TREATMENT

Keep weed-free (hoe) until the plants get to a height where they will not have to compete with the weeds for light and the weeds stop growing in the autumn. Mulch in summer to help retain moisture and in winter to help protect the roots from ground freeze. Watch out for slugs and snails – they love this plant.

HARVESTING

Harvest the bottom leaves whenever you are hungry. If you cut the top off, the plant will grow side shoots but they won't be as good as leaving the main shaft intact.

RIGHT: Cavolo Nero is a traditional Tuscan plant. It is wonderful prepared on its own or added to soups and stews to give colour and vitamins in the winter months when other fresh vegetables are scarce.

STORAGE

Cabbage can be frozen once it is in a stew or soup or otherwise pickled (which isn't worth it as you should be eating it through the winter months fresh).

EXPECTED YIELD

You should be able to feed a family of four on 4 plants if you eat cabbage once a week.

VARIETIES

In my opinion there is only one Italian cabbage worth growing: 'Cavolo Nero di Toscana'.

cauliflower

cavolfiore

Cauliflower is another superb winter brassica but is a little trickier to grow than its brother broccoli. Typically, my favourite vegetable is notoriously difficult to grow but because it is so beloved I have kept at it and had a little success. For the new gardener I would advise you to look on cauliflower as a challenge but not to rely on it as a 'crop' until you've sussed out your soil and local weather. Cauliflower are not a great container or small garden crop – grow something more reliable on your precious plot.

SOIL/SITE

Sheltered site in sun or partial shade, fertilized soil slightly alkaline (make sure you have added a little lime). Dig in some ash too as you prepare the site. Don't over-manure them with fresh manure – too much nitrogen stunts their growth.

SOWING

Sow in drills or pots from late spring (as a rule, but varieties vary so follow the instructions on the packet). Plant out about 60cm (24in) apart when the seedlings have five or six good leaves.

TREATMENT

Keep weed-free (hoe) until the plants get to a height where they will not have to compete with the weeds for light. Mulch in summer to help retain moisture. Watch out for slugs and snails. Protect the curds (flowers) from weather by bending some leaves over them. When it starts to get frosty I recommend

covering the crop with fleece – extreme frost will damage the curds and make them inedible.

HARVESTING

Depending on what you have planted (autumn, winter or spring varieties), you can harvest from early autumn to late winter. They are ready when the curds are exposed. Don't leave it too long to harvest them after the curds have shown… I tend to wait for them to grow bigger (I always put them in on too rich soil so they end up small) and consequently they end up slightly mouldy and only fit for the rabbit (but even our rabbit got fed up with cauliflower last year!).

STORAGE

Cauliflower doesn't store well in the freezer – it's grown as a winter crop so you shouldn't need to store it anyway.

EXPECTED YIELD

You should get 1 large head per plant.

VARIETIES

'Romanesco' is great for a change to northern varieties because the curds are green and form in spikes. For a more traditional-looking cauliflower, try 'Marzatico', which is good in the cold.

pasta with broccoli
orecchiette ai broccoli

Orecchiette means 'little ears', which is what this pasta looks like. It is the perfect shape for picking up the sauce and the lumps of broccoli. Broccoli is a staple vegetable throughout the winter and this recipe makes it a little bit special.

SERVES SIX
2 medium-sized broccoli heads, cut into florets, stems sliced
600g (1lb 5oz) orecchiete pasta (preferably fresh)
big glug of olive oil
2 garlic cloves, peeled and diced
3 large tbsp mascarpone cheese
salt and freshly ground black pepper
freshly grated Parmesan to serve

Cook all the broccoli in boiling water until it's al dente. Fish out with a slotted spoon and put on one side. Place the pasta in the broccoli water and add more water if necessary. Cook until al dente.

Meanwhile, heat the olive oil in in a large pan over a low heat and cook the garlic for a few minutes. Add the broccoli and cook on low for a few moments to let the flavours infuse. Add the mascarpone and melt everything together.

Drain the pasta once cooked and put into the pan with the broccoli mixture. Gently cover the pasta with the broccoli sauce, season to taste and serve immediately with heaps of freshly grated Parmesan.

rustic seasonal soup

ribollita rustica

Ribollita literally means 're-boiled' and is the traditional soup that would sit on the fireside and get added to and eaten from on a daily basis. *Ribollita* is one of the Tuscan staple foods throughout winter because it utilises all the leftovers plus dried beans and the nutritionally valuable cabbage.

Serve this on a frosty winter's day and watch everyone thaw out instantly.

The vegetables that go into the soup are just a suggestion – use anything you have from the garden. The things that make it 'ribollita' are the beans, the bread and the cabbage. And that it's boiled a lot.

SERVES SIX

300g dried beans, soaked overnight

olive oil

2 small red onions, peeled and diced

2 carrots, peeled and diced

3 sticks of celery, trimmed and diced

6 large plum tomatoes, roughly
    chopped

3 garlic cloves, peeled

1 small potato, peeled and diced

1 bay leaf

pinch of dried red chillies or powder

400g (14oz) cavolo nero, leaves and
    stalks finely sliced

1 litre (1¾ pints) vegetable or
    chicken stock

2 large handfuls of good quality stale
    bread, torn into chunks

salt and freshly ground black pepper

extra virgin olive oil to serve

Drain the beans, then put them in a large pan, cover with water and simmer for 30–40 minutes until they are soft. Top up the water so that it is always covering the beans by not more than 2cm (3/4in). Drain and rinse the beans.

Meanwhile, heat 2 tbsp of oil in a large pan and cook the onions gently for 3–4 minutes until soft. Add the carrots and celery and sweat until soft, then add the tomatoes and garlic and cook for 3–4 minutes. Add the potato, bay leaf, chillies, cabbage, stock and beans, cover and simmer gently for 15 minutes. At this point the cabbage should have decreased in volume. Add some water if the soup is looking too thick.

Stir in the bread, cover and simmer for a further 30 minutes. You are aiming for a chunky, thick soup so if it's watery take the lid off and boil some of the liquid away, or if it's too thick add some water or tomato juice. Season to taste. To serve, spoon into bowls and swirl a good slurp of excellent olive oil on the top.

cauliflower and pasta bake

cavolfiore e cuocere in forno pasta

This staple Casa del Sole dish is something I normally serve to the family on a weekday night when I need to have dinner prepared in advance. You can jazz it up a bit by adding some thin strips of Parma ham to the top just before you pop it in the oven. However, it's a great dish for vegetarians without this.

SERVES SIX–SEVEN

700g pasta shapes (any type)

1 large cauliflower, cut into florets

1 medium onion, peeled and diced

1 garlic clove, peeled and crushed

splash of olive oil

750ml (1¼ pints) béchamel sauce

freshly grated nutmeg

125g (4oz) freshly grated Parmesan

Put the pasta on the hob to cook.

Cook the cauliflower with a dash of water in the microwave on full power for about 7 minutes – until all stalks are soft when prodded with a knife blade. Place in a very large baking dish. Cook the onion and garlic with a splash of good olive oil in the microwave on full power for 2 minutes.

Meanwhile, drain the pasta when it is 'al dente' and pour over the cauliflower, then mix in the onions and garlic. Pour the sauce (hot or cold – no matter) over the vegetables and pasta, sprinkle on the Parmesan and grate a little nutmeg over too. You can now store this dish in the fridge or cook it right away.

Place in an oven preheated to 175°C (350°F) until the cheese has melted and the top is golden.

asparagus

asparagi

Asparagus requires it's own bed for the whole year so you have to have a decent sized plot to be able to grow it. Once planted it needs little maintenance and will be amongst your first greens to grow in the new year. It's easy to grow and fabulous to eat: definitely worth planting.

SOIL/SITE

Any soil that is well drained and in full sun. Work the soil well and remove as many of the weeds as you can because this will be the last time you can dig the bed.

SOWING

Buy the 'crowns' (roots – which look like a stringy mop head) early spring and plant them so that the crown is about 2cm (¾in) under the surface. If the roots are really long (and the best ones are) you can lay them sideways rather than digging a hole a foot deep.

TREATMENT

Make sure the newly planted roots do not dry out. Water when very dry. Cut the fronds back in Autumn to about 5cm (2in) height. Cover in winter with compost or manure. They also like the minerals from ash.

AFTERCARE

Watch out for weeds and slugs

HARVESTING

For the first year leave the asparagus uncut. In the second year you can cut a few of the first shoots. From the third year on you can harvest as much as you like. The more you cut the more will grow. Harvest when the spears shoot up but before the ends have unfurled their foliage or the flesh will be woody and tough.

STORAGE

You can pickle asparagus or freeze it (though on defrosting it will be limp and only really good for soup). It's best eaten completely fresh.

EXPECTED YIELD

2kg (4 ½lb) per sq m (sq yd)

asparagus with poached eggs

asparagi con uova in camicia

Asparagus arrives almost overnight in the late spring. It grows so fast that you have to keep a keen eye on the garden or the spears will have grown too long and go to seed. As soon as you see the asparagus peeping above ground get ready to enjoy it, as the season can be quite short. Home-grown asparagus is one of the most heavenly tastes in the world, hence this simple and classic recipe.

SERVES SIX

1kg (2¼lb) asparagus

6 eggs

butter

sea salt and freshly ground black pepper

Parmesan to serve

Clean the asparagus and chop off the woody bottoms, leaving just the green stems.

Boil for 7–10 minutes until tender but still slightly crunchy. Remove from the water, pat dry with a teatowel and stack in neat piles on six plates.

Meanwhile, bring a pan of water to the boil for the poached eggs. Once it is simmering break the eggs into the water. There is no mystery to good poached eggs – it is all in the freshness of the eggs. Eggs over four days old will not hold together well in the water, so use a poacher.

Once the egg whites are opaque and hard, remove from the water using a slotted spoon and place one egg on each pile of asparagus.

Season, then top with a little butter and shave the Parmesan over each plate to taste.

onions, garlic & leeks

onions

cipolle

So many dishes need onions that they really are a must in your kitchen garden. I have grown them in all three different ways, from sets, from seed and from immature seedlings, and all ways worked well but they are quite high maintenance from a weeding point of view because their spindly leaves are no good at thwarting the competing weeds.

SOIL/SITE

Sunny site with moist, well-drained and well-dug soil that has been tamped down well – onions need firm soil or they keel over because their roots aren't all that strong.

SOWING

Outside, sow in drills 2cm (¾in) deep and about 15cm (6in) apart from early spring. From seed, my preference is to sow in a seed tray and then transplant when the seedlings are ready to about 10cm (4in) apart. The closer they are to each other the smaller the onions will be. If sowing sets (heat-treated baby onions) just pop them in the ground 1cm (½in) under the surface, in drills 10cm (4in) apart with a 10cm (4in) space between them and watch them grow (this is the easiest but most expensive way of growing onions). Sometimes, if I have forgotten to plant the onion seedlings early enough I plant the little seedlings that you can get at market gardens in spring. These are treated just like the seedlings you grow yourself but save you the hassle of bringing on seeds.

TREATMENT

Onions like water if it is very dry but don't worry about them too much unless there is a drought.

AFTERCARE

Keep them weeded well or your onion crop will suffer in size. Some people turn onion leaves over on themselves and tie them up with an elastic band, the idea being that the nutrients go back into the bulb. I'm not convinced; happy plants are normally natural plants.

HARVESTING

Early plantings should be ready in mid- to late summer and successive plantings should be ready after this. Harvest before the first serious frosts. They are ready for pulling when the leaves start to die off but if you need an onion before that happens you can always pull some up early.

STORAGE

Onions keep well if stored in a dry, dark place. I like to plait mine into ropes and hang them up in a cupboard. You can also store them in a pair of old tights – pop them down the legs and make a knot after each one. They don't like being stored on the ground or in piles. Small onions store better than large ones and as soon as it gets to be springtime they detect it and start sprouting, so try to have used them all up by late winter.

EXPECTED YIELD

10 per 1m (3ft) row.

garlic

aglio

This is one Italian plant that thrives in cooler temperatures. In fact, it positively needs a good frost. A fairly easy plant to rear and it's indispensable in the kitchen. Plant and treat exactly as onion sets.

SOWING

Make two sowings, one in late winter and one in late autumn for year-round fresh garlic.

VARIETIES

Italian Purple, 'Rosso di Sulmona'.

leeks

porri

Leeks are not only easy to grow, they grow in winter just when you need them. They resist frost, so they don't need any storing – just pick them as you need them. What a thoughtful little veg. I put them into the bed that I have just taken my early potatoes out of.

SOIL/SITE

They grow pretty much anywhere, so pop them in a bed that nothing else likes much and they will do fine.

SOWING

Sow in drills or cells from early spring (indoors from mid-winter). Plant out when they have four or five good leaves. They need to go in about 15cm (6in) apart and if you want them to get lovely white trunks you need to use a dibber (fat, pointed stick used for making holes) and make a widened hole about 20cm (8in) deep, then drop the plant in the hole and water in.

RIGHT: Leeks grow big. They need pulling up and using before the outer skin starts to split, or the whole vegetable will be spoilt.

TREATMENT

Leeks only need attention if there is a drought, when they appreciate a bit of water.

AFTERCARE

Like onions, keep them weed-free and bank soil up the stems as they grow to encourage a long, white trunk.

HARVESTING

Pick when needed: leeks taste wonderful when they are small but you can normally leave them in the ground until you need them, into the next spring.

STORAGE

If you have to pull them because you need the space, you can 'heel them in' (see p.xx) and they will keep until you need them.

EXPECTED YIELD

10 per 1m (3ft) row.

VARIETIES

'Gigante d'inverno' – the perfect winter leek.

leek and potato soup

porri e patate

This hearty soup is a perfect winter warmer. Leeks are normally ready from late autumn and then throughout the winter. Potatoes are easily stored and you should be aiming to use them up through winter.

SERVES SIX

4 large potatoes, peeled and diced

4 large leeks, trimmed but still with
   a bit of greenery

2 vegetable stock cubes

salt and freshly ground black pepper

1 tbsp mascarpone or 2–3 tbsp of
   single cream

Put the potatoes in a pan of water. Slice the leeks into thinish rounds and add to the pan, then crumble in the stock cubes. There should be sufficient water in the pan to cover the vegetables but no more than this. Cover with the lid (this is important to conserve all the flavours of the stew and to avoid steaming the whole house) and boil until the potatoes are soft.

Take a potato masher and mash the soup into the consistency you prefer (I like mine smooth but some like it chunky). Season to taste. If the soup is slightly too thick add some more water or a little milk. If too thin then boil for another few minutes without the pan lid. Add the mascarpone or cream (don't boil it after the cheese or cream is added – it can curdle and look nasty, although it will still taste fine) and stir to combine, then serve.

mussels with baby leeks

cozze con porri bambino

Little leeks are often available in the spring, but they are normally available in late summer and autumn, so whenever you harvest them, use for this fresh and tasty dish.

SERVES FOUR

olive oil

2 tbsp butter

5 baby leeks, cleaned

3 garlic cloves, peeled and finely sliced

90ml (3fl oz) Marsala, sherry, or white wine (optional)

150ml (¼ pint) cream

1.1kg (2½lb) mussels, cleaned and debearded

handful of parsley, roughly chopped

salt and freshly ground black pepper

Heat a good slug of olive oil and the butter in a large pan, then slowly fry the leeks and garlic for about 5 minutes, when they should be very soft and sweet to taste.

Pour in the alcohol, turn up the heat and simmer for 1 minute. Add the cream, bring back to the boil and pop in the mussels. Cover the pan and boil for 5–7 minutes until all the mussels have opened, shaking the pan occasionally.

The mussels are cooked when they have opened up – discard any mussels that remain closed. Stir in the parsley and season with salt and pepper. Serve with big hunks of crusty bread.

RIGHT: Parsley is easy to grow and indispensable in the Italian kitchen. The most commonly used type is flatleaf.

leek tart

torta di porri

Vegetable tarts are often served at lunchtime. Traditionally they are fairly shallow but usually I have so much of the vegetable ingredient that I tend to make them a little deeper than I should. You can use this base recipe for any vegetable (or mixed vegetable) tart.

TO MAKE A DEEP
20cm (8in) FLAN

FOR THE PASTRY:

125g (4oz) salted butter

250g (9oz) plain flour

1 tbsp olive oil

5 tbsp ice-cold water

FOR THE FILLING:

about 3 leeks, trimmed, rinsed and
    sliced into rounds

1 garlic clove, peeled and crushed

slurp of olive oil

100ml (3½fl oz) white wine (optional)

3 eggs

250ml (9fl oz) single cream

salt and freshly ground black pepper
    nutmeg

150g (5oz) freshly grated Parmesan
    cheese (or any old cheese if you
    haven't got Parmesan handy)

Dice the butter and rub into the flour with your fingertips until it resembles breadcrumbs (and you can't find any more lumps of butter in the mix). Using a fork, stir in the olive oil and cold water. Mix with the fork until the mixture holds together. If necessary, get your hands in there and mould the dough (*pasta* in Italian). Wrap in clingfilm and put in the fridge to rest for about half an hour.

Preheat the oven to 180°C (350°F).

Sauté the leeks lightly in the olive oil until they start to go opaque. Then add the wine and cook until most of it has evaporated. If you are not cooking with alcohol, add about ½ a cup of water and cook on a low flame for another 10 minutes until there is only a little leek juice left in the pan. Set aside for a moment.

Roll out the pastry and line your tin/flan dish. Prick the flat surface of the pastry so that it doesn't rise, and place in the heated oven for 10–15 minutes.

Meanwhile, whisk the eggs and cream in a bowl. Season, add the leeks and mix thoroughly.

Remove the pastry case from the oven and make sure it is cooked sufficiently for the sides not to collapse when you move it. Pour the leek mixture into the partially cooked pastry case. Grate the nutmeg generously on top and then sprinkle on the Parmesan.

Cook for a further 20–30 minutes until the top is golden. Leave to cool before removing from the tin. Serve warm or at room temperature and don't keep for too long or the pastry base will go soggy.

baked spring onions
cipolline al forno

Baby onions are delightful and in abundance when you are thinning your onion rows.

SERVES SIX

1kg (2¼lb) baby onions, peeled

2 bay leaves

olive oil

scant tbsp unrefined brown sugar

sea salt

Preheat the oven to 200°C (400°F). Place the onions and bay leaves in a pan of boiling water and boil for about 10 minutes, then drain and dry well with a towel.

Add a good slug of olive oil to a large baking tray and heat in the oven. Put the onions in and slosh around to ensure all are coated in oil. Sprinkle the sugar and salt over the onions and roast for 20 minutes, shaking halfway through to avoid burning the undersides.

beans & peas

beans

fagioli verdi

My Dad used to grow string beans and I've never liked 'em! However, I love French beans and the best thing about this type is that the more you pick, the more you get. The dwarf varieties are perfect for container gardening. I also like to grow a couple of other varieties especially for drying and using in soups in winter. For this I plant 'Red Mexican' and 'Cannellini', not normally from a proper seed shop but from the local organic food market since normal dried beans should germinate just as well as those bought in packets and usually cost considerably less.

SOIL/SITE

Sunny site with moist, well-drained soil. They do not like acid soil, so if you have a mystery problem with your peas look into the pH of your ground: they like it limed.

SOWING

Sow in drills 3cm (1in) deep and about 5cm (2in) apart. Peas are pretty hardy, so you can sow from early spring to mid-summer with most varieties and use 'early' ones in late winter. If sowing indoors, plant two peas in every cell and thin out the weakest.

TREATMENT

'Pole' varieties climb and will need some sturdy bamboo wigwams to climb up. If you are gardening in a small space then climb them up a trellis against a wall and plant a few bush varieties at the bottom to optimize space.

AFTERCARE

Keep the weeds down until the plants are large enough to mulch. They really benefit from a good mulching through summer, as it keeps the soil cool and moist.

HARVESTING

Whenever the beans are looking plump pick them. The more you pick the more will grow. If you leave beans on the plant it will not produce more beans. If leaving the peas for drying then leave them on the plant until it browns, then pick the pods, use your thumb to chase out the beans and store them away somewhere cool and dry. Once finished, leave the roots in the ground and dig them in as a fertilizer (beans store nitrogen in their roots, which is a potent fertilizer).

STORAGE

French beans should be eaten fresh within a few hours of picking. You can freeze them but they take up a lot of room and never taste as good as the fresh ones.

EXPECTED YIELD
1kg (2¼lb) per 1m (3ft) row.

LEFT: Black fly infesting a young bean plant. The only organic way to get rid of them is to introduce ladybirds, who will eat them. In this case, however, the problem is slight and they will not affect the crop.

peas

piselli

Peas are amazing when fresh from the garden – sweet and delicious. Even frozen peas are a poor alternative to fresh from the garden, as the sugar in the pea starts to turn to starch within a few hours of picking.

SOIL/SITE
Sunny site with moist, well-drained soil. They do not like acid soil, so if you have a mystery problem with your peas look into the pH of your ground.

SOWING
Sow in drills 3cm (1in) deep and 5cm (2in) apart. Peas are pretty hardy, so sow from early spring to mid-summer with most varieties and use 'early' ones in late winter. If sowing indoors, plant two peas in every cell and thin out the weakest.

TREATMENT
The 'wrinkled' variety (when it grows it's normal) is pretty hardy but it's best if you can protect them with fleece. Mice and birds love peas, so if you don't have a cat, you might need to put netting over the plants or have a bird scarer. Mice steal the peas when they are sown – rinse the seeds with paraffin (doesn't have any affect on the germinating pea) or sow in pots and plant out as seedlings.

AFTERCARE
Keep the weeds down until the peas are large enough to mulch. Peas benefit from a good mulching through summer, as it keeps the soil cool and moist. Most peas will need support for their climbing habit. Stick a few branches in among your plants or make a bamboo wigwam with netting for them to climb up.

HARVESTING
Whenever the pea pods are looking rounded and full, pick them. The more you pick the more will grow. If you want to dry the peas, leave them on the plant until it browns. Once harvested, remember to leave the pea roots in the ground and later dig them in (pea roots are a great source of the nutrient nitrogen).

STORAGE
Peas should be eaten fresh within a few hours of picking. If you want to store them, drying is the most practical way (super in stews and for making mushy peas). You can freeze them but never taste as good as the fresh ones.

EXPECTED YIELD
1.25kg (2¾lb) per 1m (3ft) row.

sweetcorn

granturco dolce

In all honesty, Italians don't eat sweetcorn unless it's made into polenta. This is because it's difficult to grow tender corn in such a hot, dry climate. Sweetcorn is divine when eaten straight from the garden but it doesn't store too well. As soon as the cob is picked the soluble sugars start to turn to starch, so if you eat it soon after picking it is delicious but the longer it is stored the less appealing its taste. Sweetcorn is simply corn that is picked early. You can leave the cob on the plant to dry out if you want to make cornflour out of it or save it to feed to your chickens or rabbits. I grow more for the livestock than I do for the family.

SOIL/SITE
Full sun, richly fertilized soil, slightly acid out of preference.

SOWING
Sow outside from late spring (as soon as the weather starts to warm up) about 15cm (6in) apart. Sow in blocks since this plant is self-pollinating by wind and needs to be near other corn plants for fertilization. You can start the corn off earlier in pots inside or under plastic – I do this because otherwise I the slugs and mice have a feast. Or you can plant sweetcorn in containers and they do really well. Put something like spinach at their base to optimize planting space.

TREATMENT
Keep weed-free until the plants reach a height where they don't have to compete with weeds for light. Mulch in summer. Watch out for slugs and snails.

HARVESTING
Pick the cobs whilst they are young and tasty. They are ready when the corn has gone yellow. If you leave them on the plant for another week they will start to be less sweet. In this case leave them there to dry and save them for your chickens.

STORAGE
Fresh sweetcorn don't store well, so eat straight away. As dried corn (maize) they store well – dry out on the plant or hang to air-dry then put the whole cobs in bags and keep dry, or scrape the corn off the cob and store in a container.

EXPECTED YIELD
You should get 2 cobs per plant.

VARIETIES
'Zuccherino' is one of the most flavourful for corn eaten straight from the garden. 'Sisred' is good for making polenta.

broad beans with sheep cheese

baccelli e pecorino

This quintessential simple summer dish uses the tastiest of ingredients fresh from the garden. Beans start to lose their sweetness once picked, so to pick and eat within a few hours is the aim of this dish.

SERVES SIX

3kg (6½lb) broad beans (fava or
    baccelli), podded and shelled to
    give 600g (1lb 5oz)
400g (14oz) pecorino cheese, cubed or
    crumbled
slurp of olive oil
1 tbsp white wine vinegar
salt and freshly ground black pepper

Put the beans, cheese, oil and vinegar in a large bowl and toss together. Season with a little salt and pepper and serve.

bean and cabbage soup

ribollita

This is a thick cabbage stew, full of winter goodness and vitamins. At a time of year when there isn't much produce in the garden *ribollita* saves the day. It is also a traditional Tuscan 'fast food' since it is normally cooked in large quantities over the fire and then eaten, reheated, over the next few days. Traditionally it is eaten at olive harvest time when every pair of hands is required in the groves and no time is spared for cooking.

SERVES SIX

300g (11oz) cannellini or other dried
    white beans
2 onions, peeled and diced
1 large head of cavolo nero (or any
    other dark-leafed cabbage), cut
    into strips
500ml (18fl oz) passata or similar
    preserved tomatoes
salt and freshly ground black pepper
250g (9oz) stale country-style white
    bread, sliced
large slurp of chilli-infused olive oil

Soak the beans overnight in cold water (you can cook them from dried but it takes so much longer). Drain the soaked beans, put into a pan and cover generously with water. Put the lid on and bring to the boil. The beans need to be softened and this could take anything up to 60 minutes, depending on what sort of bean you have used. Make sure that the beans are always covered with water to avoid burning them.

Once the beans start to soften add the vegetables and passata. Season with lots of salt and black pepper to taste, then simmer until the beans are cooked through. The soup should be a thick consistency, so if it's too runny, boil without the lid for a short time, or add water if it is too thick. Once the soup is cooked, take it off the heat and adjust the seasoning. If you like a creamy soup, take a potato masher to it, otherwise leave it chunky.

To serve, place one trencher of bread in the bottom of a bowl and ladle in the soup, then drizzle a little olive oil over the top. Leave to stand for a moment before serving to let the soup soak into the bread.

beans in tomato sauce

fagioli all'uccelletto

Beans are such a staple of the Italian diet because they are easy to grow and store well through the year. This hearty dish is made in summer with fresh tomatoes and through winter with preserved tomatoes. The recipe was given to me by the owner and chef at 'da Sandrino', a super little restaurant in the heart of the Sorana valley bean-producing area. Attilio and his ancestors have run the restaurant since the 1800s. Sorana beans are famous throughout Italy for their delicate flavour and extremely thin skins.

SERVES SIX

500g (1lb 2oz) cannellini beans (or preferably Sorana beans if you can get hold of them)

slurp of olive oil

300g (11oz) tomatoes, roughly chopped

2 garlic cloves, peeled and minced

5 large leaves of fresh sage

10 whole peppercorns

salt and freshly ground black pepper

Soak the beans overnight in cold water. Drain the soaked beans, put into a pan and cover generously with water. Put the lid on, bring to the boil and cook for about 40 minutes – maybe more – until soft.

Heat the olive oil in a pan, add the tomatoes, garlic, sage and peppercorns and cook for about 5 minutes until thickened. Season this sauce with salt and pepper.

Strain the beans and keep ½ cup of the water. Add the beans and reserved water to the tomato sauce and stir over a gentle heat. The sauce shouldn't be too runny – cook a little longer to evaporate if necessary.

spaghetti carbonara with peas

carbonara primavera

This superb dish is served in a restaurant near to us in the village of Montecarlo; it has fantastic views from the terrace and the wines from this area are outstanding.

SERVES FIVE/SIX

salt and freshly ground black pepper

500g (1lb 2oz) spaghetti

400ml (14fl oz) milk

4 eggs

1 slurp of olive oil

1 small onion, peeled and diced

1 large handful of fresh podded peas

250g (9oz) pancetta (if you can't find this, then fried bacon will do – I use smoked), diced

75g (3oz) freshly grated Parmesan

Bring a large pan of salted water to the boil and add the spaghetti, cook for about 6 minutes (it normally tells you how long on the packet) until *al dente*. Whisk together the milk and eggs, add a pinch of salt and a few grinds of black pepper.

Drain the pasta when it is ready and leave in the colander for a moment. In the same large pan, heat a slurp of olive oil and fry the onion until opaque (a couple of minutes) but not browned. Put the pasta back in the pan and keep the heat on low. Add all the other ingredients and stir until the sauce thickens.

Serve immediately.

potato

di patate

Potatoes aren't prevalent in Italian cookery like they are in many northern countries. However, there are some very Italian dishes that rely on them. Gnocchi is one. And potatoes are so easy to grow. If you choose an early variety you can pick the potatoes and later plant something else in the same ground. I normally plant leeks for winter in the furrows left from my early potatoes.

There is little more satisfying than lifting your potato crop and rootling through the dirt for treasure. Potatoes are easy to grow, store well and, although they take up a lot of room, even the container gardener can have a go – simply half fill a large container with well-rotted manure, put in as many potatoes as you have room for (see Sowing, below) and cover with more compost.

SOIL/SITE

Heavily mucked soil. Plant potatoes on ground that is heavy and needs work – the roots will help break up the soil, so you won't break your back digging.

SOWING

Chit potatoes about six weeks before planting to give them a head start: place the seed potatoes, the ends with most 'eyes' on uppermost, on a windowsill to encourage them to sprout. Pop them in the ground when the sprouts are 3cm (1in) or so long. Plant early varieties as soon as the last frost is over. If you make a mistake and put them in when there is a frost, don't worry too much unless there are shoots poking up already – in which case cover them with fleece until it gets a little warmer.

Dig a channel and put the tubers in about 35cm (14in) apart. Cover with the soil and then a huge amount of rotted manure. Potato varieties come in 'early' or 'maincrop'. I always only plant earlies because you can plant these in early spring (or earlier if you protect them with plastic or a polytunnel), lift them at the end of spring and still have loads of time to plant something else on that plot. Maincrop take up a lot of room for a long time and are also susceptible to blight in the warmer, muggy summer conditions.

TREATMENT

When the sprouts start showing above ground 'earth up' (cover the sprouts so that you end up with a raised mound). This protects the new shoots from the cold, encourages more root growth (hence more potatoes) and makes sure that your crop is well buried and you won't get green potatoes (which are poisonous and occur when a normal potato is subjected to light whilst it is still attached to the root system).

AFTERCARE

Weed until the plants are large enough to smother the weeds themselves. Keep the ground moist in dry times – potatoes will crack if not given enough water.

HARVESTING

You can harvest your potatoes whenever you like, lift them early for little potatoes and later for biggies. Some traditions have you lifting them as soon as the flowers have died, but I have found that as long as your site is free-draining you can leave them in as long as you need to.

STORAGE

Potatoes can be clamped (see p.xx) successfully through winter. Once out of the clamp, store in a dark dry place.

EXPECTED YIELD

1kg (2¼lb)+ per plant.

VARIETIES

As far as Italians are concerned, 'Primura' (which is an early potato) is one of the best. However, getting seed potatoes of an Italian variety can be tricky, so just go for any floury variety if you are growing them to make gnocchi (as you need particularly dry potatoes for this).

gnocchi with pesto

gnocchi al pesto

Gnocchi is a really filling dish and is always a *primo piatto* because it fills you up nicely before you move onto the meat course. It's not as difficult to cook as you might think.

The secret to good gnocchi is to keep the potatoes as dry as possible, which is why I bake mine (and I like to eat the skins with some butter on them as I'm cooking) but you can also boil them in their skins and peel them once cooked to reduce wateriness.

SERVES SIX

1kg (2¼lb) baking potatoes

2 egg yolks

125–185g (4–6½oz) plain flour

salt

150g (5oz) pesto (see p.133)

Bake the potatoes for about 1 hour, then leave until cool enough to handle. Scoop out the flesh and mash with a fork, trying to keep it as fluffy as possible – a potato ricer is perfect for this job. Mix in the egg yolks and then gradually stir in the flour. Once a loose dough forms stop adding flour and transfer to a floured worksurface. Knead gently and carry on adding flour until you have a soft dough that is damp but not sticky to touch.

Divide the dough into about six portions and roll each out into a rope the thickness you require for your gnocchi. Using a sharp knife, chop the gnocchi into small cubes, then press a fork into the top of each piece to give it ridges.

To cook the gnocchi, bring a large pan of salted water to the boil. Drop in the gnocchi in batches, so that they are not too crowded in the pan, and cook for about 2 minutes – when they rise to the surface they are done. Drain and immediately add the pesto sauce, stir gently to coat all the gnocchi and serve.

potato bake from the north of Italy

torta di patate alta adige

Southern Italy sticks to pasta because this is what is readily available, northern Italy flirts with potatoes, rice and polenta because these things grow in the cooler and more temperate north. The northern Alta Adige region is famous also for its dairy produce. This dish is normally served as a *primi* but can also be served as a *contorni* dish (side dish to the *secondi* course).

SERVES FOUR

olive oil

1kg (2¼lb) potatoes (peeled or
    unpeeled), thinly sliced

2 garlic cloves, peeled and thinly sliced

1 small onion, peeled and thinly sliced

100g (3½oz) mozzarella, sliced

50g (2oz) salted butter

2 tbsp single cream

50g (2oz) freshly grated Parmesan

Preheat the oven to 170°C (325°F). Grease an ovenproof dish with a little olive oil.

Layer the potatoes, garlic, onion and mozzarella in the dish. Melt the butter and pour it over the potatoes, then spoon on the cream. Finally, sprinkle on the Parmesan and cook for about 50 minutes until the potato is cooked right the way though. Leave to stand for 10 minutes before serving.

herby roast potatoes

patate arrosto erbaceo

These super potatoes are delicious and an excellent healthy alternative to standard roast potatoes. You can use dried herbs if you don't have them fresh but it isn't quite the same.

SERVES SIX

large slurp of olive oil

1 sprig fresh rosemary

bunch of fresh thyme

bunch of fresh sage

6 medium-sized potatoes, unpeeled and
    cut into cubes

generous 1 tsp rock salt

Heat the oven to 225°C (425°F). Put a large glug of oil in a big roasting dish and heat in the oven for 5 minutes.

Meanwhile, chop all the herbs into fairly small pieces.

Place the potatoes in the pan (watch out for the oil spitting at you), add the salt and all the herbs.

Roast for about 35 minutes, until browned and cooked through.

other vegetables

carrot

carota

In my experience, carrots are tricky to grow. First I started out with a heavy, clay soil that they didn't like to send their roots down into. The next year I had improved my soil with so much manure that the carrots went mad and forked off all over the place. One day I'll have the perfect, deep loam needed, but until then I'll only grow the snub-rooted varieties, which don't get upset by shallow, stony or over-fertile soil.

SOIL/SITE

Deep, stone-free, sandy soil.

SOWING

Carrots must be sown directly into the bed or container they are going to grow in. You can't sow carrots in pots because they don't like moving and it just attracts the carrot root fly, so sow *in situ* 2cm (¾in) deep in rows 30cm (12in) apart in late spring to summer. If you sow every 2–3 weeks you will have a constant supply of carrots through autumn. Sow thinly so that you don't have to do too much thinning (the carrot fly sniffs out the scent of crushed carrot leaves during the thinning process). I intersow my carrots with my onions and have not yet had a carrot root fly problem. You can grow carrots in pots – they grow really well in compost with a bit of sand added but take up a lot of space. You can intercrop them with onions, however, to keep the carrot fly away and they don't compete for each other's space.

TREATMENT

Thin the seedlings out to 10cm (4in) apart to start with. As they grow, take out every other baby carrot (good in salads) so that the distance is then 20cm (8in), which is good for growing big carrots for storing in winter.

AFTERCARE

Weed, weed and more weeding. Hoe if your plants are spaced widely enough for it not to disturb the roots, but really you are best off doing it by hand.

HARVESTING

Pull up young carrots when you want them for your dinner. For your main crop (the ones you want to store), leave until they are nice and big, normally late autumn before the first frosts. Dig them out with a fork to avoid damaging the root. Try to harvest on a dry day and leave them on top of the soil to dry off a bit. Brush them clean with your hand but never wash them – they won't keep more than a week if they get wet. Trim off the green tops with a knife and compost them.

STORAGE

Carrots are best stored in sand. I put a layer of carrots (trying not to let them touch each other) in a box and then cover them with sand and repeat. They store all winter like this.

EXPECTED YIELD

1kg (2¼lb)+ per 1m (3ft) row.

VARIETIES

'Nantese di Chioggia' is a good Italian variety. If you have bad soil or limited space (i.e. in pots) try the unimaginatively named 'Mini Round'.

LEFT: It is really important to label your seeds well, as many plants look similar when they are small.

aubergines
melanzane

Aubergines are beautiful to look at and, when home grown, are tasty too. A real Mediterranean vegetable, they are not too difficult to grow as long as they are given a sheltered, sunny spot. A very good container plant since it grows high and you can plant something else below it (I put strawberries under mine).

SOIL/SITE

A sheltered, sunny spot that has been manured well but not recently. They also grow well in grow-bags or fair-sized, deep pots.

SOWING

Don't try to sow outdoors – sow in cells early to mid-spring. Keep in a greenhouse or conservatory (it has to be around 20°C) until the plants are about 30cm (12in) high. Then you can plant outside, preferably against a south-facing wall, but they are more likely to crop well in a greenhouse or on a window ledge.

TREATMENT

Keep moist but not wet. Mulch through summer and don't let them grow too high – pinch out the end sprigs when the plant gets to be 70–80cm (28–30in) high.

AFTERCARE

Weed until the plants are large enough to mulch. Ensure the fruits aren't too heavy for the branches – if necessary, support them with twigs. Don't let the fruit come into contact with anything or the skin will be scarred.

HARVESTING

When the fruits are glossy purple they are ready to harvest. Don't try to pull them off the stalks – you'll damage the plant, instead use secateurs.

STORAGE

The only way of storing them is to slice and either grill them and put them under oil or dry them (good for using in winter lasagne). Otherwise, just eat them fresh – it's unlikely that you will have too many.

ESTIMATED YIELD

Normally about 5 aubergines per plant.

VARIETY

The round 'Kamo' variety is popular in Italy, as is 'Violetta lunga', which is a more traditional-looking aubergine.

fennel

finocchio

Fennel is greatly loved by Italians. Not only does it taste wonderful, but it also aids digestion and, due to some clever phytoestrogens, is great for balancing hormones. It's a wonder plant in other words. It was taken from its indigenous home in Italy all over the world by the Romans, who used it as a valuable medicinal plant more than an addition to the dinner table.

SOIL/SITE

A sheltered, sunny spot that has been manured well but not recently. They also grow well in pots.

SOWING

Sow outdoors or in seed trays in early spring and do a second sowing in early summer for a winter crop.

TREATMENT

Allow them to dry out occasionally.

AFTERCARE

Keep weed-free, as they can get choked and stunted with too many weeds. They can withstand frost but not a complete freeze – pull up if you know there is going to be a particularly cold snap.

HARVESTING

Eat at any point in growth but it's obviously better to leave them to mature into larger plants.

STORAGE

Freezes quite well for use in stews, etc., but obviously loses its wonderful crunchy texture.

EXPECTED YIELD

1kg (2¼lb)+ per 1m (3ft) row.

VARIETIES

'Florence' is the quintessential Italian variety.

aubergine with parmesan

melanzane alla parmigiana

This Italian dish uses loads of aubergine and so is normally cooked at the end of summer when the aubergines are in full fruit. 'Costoluto Fiorentino' are the best tomatoes to use in this dish.

SERVES SIX

3 large aubergines, thinly sliced

olive oil

flour for dusting the aubergine slices

2 garlic cloves, peeled and minced

salt and freshly ground black pepper

5 very ripe tomatoes, sliced

2 sprigs of fresh oregano, stalks
    removed

300g (11oz) mozarella cheese, sliced

50g (2oz) freshly grated Parmesan

Preheat the oven to 180°C (350°F). Preheat a griddle pan.

Brush the aubergine slices with oil and griddle on both sides for 2–3 minutes. Transfer to a plate and dust with flour. Some people like to salt cure their aubergine to stop it tasting bitter but I have never had a bitter aubergine from my garden. If they are ripe (when lightly squeezed they yield) they should be fine without salting.

Put the garlic in a bowl, add a slurp of oil and season well. Layer the ingredients in an ovenproof pan or dish, brushing each layer with the garlic oil. Drizzle some more olive oil over the dish, cover with foil and cook for 30 minutes. Then remove the foil, sprinkle the Parmesan over and cook for a further 30 minutes or until the top is browned and the aubergine is cooked.

fennel gratin
finocchi gratinati

This dish uses a lot of fennel, which is handy when you find that you grew too much…

SERVES SIX

6 fennel bulbs

plain flour for dusting

salt and freshly ground black pepper

375ml (13fl oz) milk

140g (4½oz) salted butter, diced

100g (3½oz) freshly grated Parmesan (or any other cheese you have handy)

nutmeg

Preheat the oven to 180°C (350°F). Steam the whole fennel for about 15 minutes or until they are soft but still firm. Then cut the fennel into quarters and sprinkle them all over with flour.

Place in an oven dish and season with salt and pepper. Add the milk and sprinkle the butter randomly over the surface. Top with the cheese and a generous grating of nutmeg. Cook for about 30 minutes or until the top is golden.

wild boar stew
spezzatino di cinghiale

Wild boar is a very Italian speciality, popular in the mountainous regions where hunting is traditional. The meat can be extremely gamey. This recipe was gifted to me by Attilio, a local chef.

SERVES SIX

1.5kg (3lb 5oz) boar meat, preferably shoulder

vinegar (optional)

large slurp of olive oil

3 whole cloves

2 bay leaves

5 juniper berries

cinnamon

500ml (18fl oz) red wine (optional)

2 onions, peeled and roughly sliced

2 garlic cloves, peeled and roughly sliced

3 carrots (or more if you have a glut), peeled and roughly sliced

1 celery stalk, roughly sliced

1 litre (1¾ pints) passata (or preserved tomatoes)

salt and freshly ground black pepper

Wash the boar meat and dice into bite-sized pieces. If it smells really strong you can rinse it in a little vinegar to take the sharpness away.

Heat a good glug of olive oil in a large pan and sauté the boar meat on all sides for a few minutes. Add the cloves, bay leaves, juniper berries and a grate or two of cinnamon and cook for another few minutes, then add the red wine if you are using it. Add the vegetables, passata and 250ml (9fl oz) water. Stir and bring to the boil, then simmer on a very low heat without a lid for about 1½ hours. Stir every now and then to prevent the bottom from burning and season with salt and pepper halfway through cooking. The stew is done when the boar meat is no longer tough.

Add more passata for more sauce, if necessary, with the stew and to keep it moist.

herbs & wild ingredients

What would Italian food be without herbs? Decorative in the garden and on the plate – the Italians often refer to them as 'drugs', in the sense that they have medicinal properties as well as being an integral part of the taste of Italy.

Herbs are ideally suited to container gardening since most like well-drained soil and don't mind drying out on the odd occasion. You can pop them in a large planter together (none of the herbs mentioned below are invasive, so will not strangle each other) and keep them outside your kitchen door for convenience. However, it's often easier to buy established plants if you want to start using them straight away.

The following herbs, excepting basil, are perennial, so you don't have to replant them every year.

oregano

origano

A margherita pizza just isn't the same without a generous dose of oregano and fresh is certainly best. This plant likes the sun and needs a little space, so don't overcrowd it in a pot. I plant mine around the bottom of my bay tree to give both room to grow.

STORAGE

In winter the plant goes a little dormant, so the leaves taste of nothing. At this time it is good to have a previously harvested crop available. Drying the leaves works quite well but the best thing to do is harvest and freeze in small containers.

bay

baia

Bay is a tree – a decidedly large tree if left to grow unchecked, so watch out. It's best to keep it contained in a pot and prune regularly to encourage new leaf growth. It will survive quite a harsh freeze if planted in the ground but if the roots freeze it will not be happy at all, so swathe potted trees in fleece at very cold times of the year or bring inside into a cool hallway or greenhouse.

STORAGE

Bay dries very well but you don't really need to bother since the leaves stay full of flavour all year, so you can pick them fresh whenever you want.

rosemary

rosmarino

Rosemary is used extensively in Italian cooking. It is hardy and should survive well outside in most places. Although it grows well in a pot, it really likes to be in the ground. It can grow into a huge bush if planted out and can also become a little straggly, so prune back well after it has flowered.

STORAGE
As with bay, you can dry it well, but why bother when you have your year-round fresh stock?

thyme

timo

Thyme is one of the herbs you find in packets of 'Italian Herb Mix'. It is great in meat sauces and I like it with roast chicken too. If looked after it will last through the winter (you may need to cover it or bring it into somewhere sheltered during very cold spells if it is in a pot). Grow a mixed container of variegated sage, oregano and thyme and it will not only look wonderful but also provide your kitchen with a great herb assortment.

sage

salvia

Sage is a wonderful Mediterranean plant. It likes sunny and dry conditions. Ideal for well-draining planters, sage will give a more intense flavour if kept dry. I like the large leaf varieties because I use a lot of sage, but for a pretty herb pot the small, variegated leaf sage is pretty and tasty.

parsley

prezzemolo

Parsley is used in all Italian cooking where there is garlic and with almost every fish dish. Flat leaf parsley is the Italian variety. Technically it is a perennial (lives throughout the year) but I find that you are very lucky, especially in cooler climates, if this is the case. You can grow it in pots and bring them inside for the winter or you can sow it every year for a new crop. The seeds take about a month to germinate and should be planted in early spring.

basil

basilico

Basil is an annual and this means that you need to plant it from seed every year. It's easy to grow but you need to start it off inside in northern climes. It is a great pot plant and will also grow happily inside on a windowsill.

SOIL/SITE

Basil needs a rich earth.

SOWING

Sow *in situ* in the garden after the last frost. In cold climates you will find that growing from seed on a windowsill or in a greenhouse from mid-spring will give you a good crop. Alternatively, you can sow under mini poly tunnels.

TREATMENT

Weed well.

AFTERCARE

Basil needs to be watered well but not kept wet.

HARVESTING

Harvest the leaves as you need them. If you pinch out the top growth you will encourage the plant to become more bushy. Pinch out the flowering heads since once it has flowered basil loses a lot of its taste.

STORAGE

Basil loses a lot of its flavour when dried so if you dry it remember you will need to use a lot of it. It can also be chopped and frozen. The best way to preserve it, however, is in pesto (see p.133).

EXPECTED YIELD

12 plants per msq (sq.yd).

VARIETIES

You can get all sorts of basil, including a lovely purple variety, but the 'Classico' is what you will find in most Italian kitchens.

rosemary pizza bread

pizza di pane al rosmarino

This is a simple yet delicious appetizer. Served in a pile in the middle of the dinner table, this is a perfect beginning to a sociable dinner and as it cooks the rosemary smells divine. In Italy, many restaurants serve this as soon as you sit down, so that you won't starve whilst you are looking though the menu! Rosemary is a perennial (doesn't die off in winter) and the needles stay full of flavour throughout the year and so you can make this bread at any time.

SERVES SIX AS
AN APPETIZER

1 batch of pizza dough (see p.81)

3 large sprigs of fresh rosemary
(dried if you don't have fresh)

plenty of good olive oil for drizzling

sea salt

Preheat the oven to as hot as it goes. Roll out the pizza dough to about 4cm (1½in) thickness and place on a tray(s). Sprinkle the rosemary over the dough and place in the oven. Depending on the temperature, the dough should be cooked in 5–8 minutes (it's done when it starts bubbling up and the edges start to brown.) Remove from the oven, place on serving plates and drizzle generous amounts of excellent olive oil over the surface. Finally, sprinkle some sea salt over and serve.

RIGHT: Work the dough hard – don't be afraid to be rough with it. This is great stress relief!

nettle risotto

risotto ortica

This is one of my favourite dishes because the nettles are free and it uses ingredients I haven't had to slave over in the garden. It also indicates the onset of spring, as nettles are one of the first things to start growing after a long winter. It's best made with new, very green shoots. Don't forget your rubber gloves!

SERVES FOUR

2 large handfuls of young nettle leaves

2.5 litres (4¼ pints) vegetable stock

50g (2oz) butter

1 onion, peeled and very finely
    chopped

400g (14oz) risotto rice

125ml (4fl oz) dry white wine (optional
    – if not using, replace with milk)

salt and freshly ground black pepper

75g (3oz) cold butter, cut into cubes

100g (3½oz) freshly grated Parmesan

salt

handful of fried sage leaves to garnish

Keeping your rubber gloves on, very finely chop the nettle leaves.

Bring the stock to a boil, then reduce to a simmer. Melt the 50g (2oz) butter in a heavy-based pan, add the onion and cook gently until translucent. Add the rice and stir to coat it in the butter and to 'toast' the grains. Once all the rice is warm, add the wine or milk. Cook until the wine evaporates and the onion and rice are nearly dry (if using milk, just simmer for a moment), then add the hot stock, a ladleful at a time, stirring constantly, and each time allowing the liquid to evaporate before adding the next ladle. After about 10 minutes, add the nettle purée and carry on cooking, adding stock as you go, until the rice is soft but still *al dente*. The risotto should be slightly drier than normal so that when you add the butter and cheese it will become the perfect consistency. Using a wooden spoon, beat in the cubed butter and Parmesan. (If the risotto is getting dry and claggy, just add a touch more milk and stir well. If it's too wet, just carry on cooking on a low heat until it comes to the correct consistency.) Season with plenty of salt, garnish with the sage leaves and serve immediately.

pesto

Pesto is such a versatile sauce, yet before I moved to Italy I had never tasted it. How can that be? It is probably because the commercial pesto you can buy in jars is good but nothing like as superb as freshly made pesto. My friends laugh at me for making my own from scratch (collecting the pine nuts and cracking them and growing lots of basil especially for my pesto) because it is such a time-consuming business but there is no denying that when it is done it's the best.

MAKES ENOUGH PASTA
FOR FOUR/SIX

1 garlic clove, peeled and roughly
    chopped
2 big handfuls of basil tips (including
    stalks), roughly chopped
25g (1oz) pine nuts
125ml (4fl oz) olive oil
8 tbsp freshly grated Parmesan

There are two ways of making pesto, the traditional and the quick.

The quick way is to bung all the ingredients in a blender and zap in short bursts until you have a rough paste.

The traditional method is worth doing if you have someone to impress because the texture is different and some argue the taste is better because of the way the ingredients are worked:

Using a large pestle and mortar, grind the garlic to a paste. Add the basil and give them some real pounding until they begin to break down. Add the pine nuts as few at a time and work them in. Then start to add the oil a little at a time and keep on grinding. Finally, scrape into a clean bowl and add the cheese, stirring in well.

This paste can be saved in jars (pop it in a jar, then seal with a layer of oil and then twist the lid on tight, which should stop any bacterial decay), frozen or used immediately.

THE Fruits

strawberries

fragole

Strawberries are enormously popular in Italy. They are one of the most valuable fruits to grow because a) they are expensive if you have to buy them, b) they absorb pesticides really easily because of their soft skin and so you really need to be eating organic ones, and c) strawberries grown at home are a million times nicer than bought ones because you can let them mature on the plant and eat them immediately.

Strawberries are a wonderful seasonal crop. They are best eaten fresh and straight from the garden. They can be preserved in jam but don't do well in the freezer.

SOIL/SITE

Strawberries like well-manured soil, so make sure they are fed well. I grow them in pots around the decking and in hanging baskets – which is a good way to keep slugs at bay – but because the kids like them so much I have established a large raised bed, which keeps them all very happy.

SOWING

To start with, buy a mixture of varieties that mature at different times to give you a longer strawberry season. Plant them about 30cm (12in) apart in spring. Later in the year the plants will throw out runners (long stalks with a tiny plant on the end). To propagate, make sure that the little plant stays attached to the mother plant but is also touching some soil. It will send roots down into the soil. Once it's established the connecting stalk will die off and you have a new strawberry plant.

TREATMENT

Keep moist but not wet. Mulch through summer. Watch out for slugs, which love to eat the fruit. As a slug deterrent and excellent fertilizer combined, liberally sprinkle ash around your strawberry plants – they love the potassium and the slugs don't like to travel over it.

AFTERCARE

Keep the weeds down until the plants are large enough to mulch.

HARVESTING

Pick when fully red. Don't leave it too long or they deteriorate on the plant.

EXPECTED YIELD

In the first year you will get about 10 strawberries per plant but this will get better over the next three years. Mature plants give more fruit but they then dip in production after about four years. Increase your crop by pinching out the runners, so the plant concentrates on flowering rather than propagating itself.

VARIETIES

For the lovely aromatic alpine strawberries choose 'Alexandria'. I also like 'Florence', which has a good resistance to mildew.

strawberry pizza
pizza di fragole

Sounds weird, doesn't it? This recipe actually became our Casa del Sole summer Pizza Party signature dish. It is absolutely divine and something very different. The strawberry season in Italy is from May to August… lucky us!

SERVES SIX

1 batch of pizza dough (see p.81)

flour to dust

a few handfuls of sugar – whichever is your favourite sort – or a little honey

1 punnet of strawberries, de-stalked and sliced but not too thinly

drizzle of olive oil

Divide the pizza dough into three balls and leave to rise for a hour or so.

Then roll out a ball into a round pizza base (it should roll out to about 15cm (6in) across but just use your judgement and don't tear it).

Preheat the oven to as hot as it will go. Unless you are lucky enough to have a proper pizza oven at home for stone baking, you need now to transfer the bases to a pizza pan (or any shallow, flat baking tray). Give this a very light sprinkling of flour before you put the dough in – this should help prevent sticking.

Sprinkle a small shower of sugar or drizzle of honey over the base. Cover the base with sliced strawberries – don't put them on more than one layer in depth and don't go too near the edge or you'll get 'run off'.

Now sprinkle the whole thing liberally with sugar (sugar works better than honey for this top coat).

Repeat for the remaining dough and then cook in the oven for about 5 minutes – you can tell when it is done because the base is crisp when you run a palette knife under it.

Remove from the oven. Drizzle olive oil lightly over the pizza, then use a pizza cutter to divide into slices and, if you're not watching your figure, serve with a dollop of mascarpone cheese. Pizza heaven!

strawberry pannacotta
pannacotta con fragole

Strawberries are the perfect fruit for pannacotta, though any fruit or a mix of fruit is also delightful.

MAKES TWELVE
SMALL RAMEKINS

15g (½oz) gelatine leaves
6 cups of single cream
250g (9oz) caster sugar
1 tsp vanilla extract
a heap of strawberries
icing sugar to dust

Soak the gelatine leaves in cold water to soften them. Heat the cream in a saucepan with the sugar and the vanilla until just at boiling point, then remove from the heat.

Take the gelatine out of the water and squeeze out. Add to the cream mixture and stir until disolved. Leave to cool for a while.

Quarter the strawberries and put as many as you need in each ramekin to cover the bottom. Then fill with the pannacotta mixture.

Chill for at least 3 hours before serving. To remove from the moulds, dip for a few moments in hot water and then invert onto a plate. Dress with a few more sliced berries and a sprinkle of icing sugar.

apples

Apples are grown in northern Italy for eating, rather than cooking. However, in the Alpine areas the crossover from Germanic countries means that there is a small tradition of strudel and apple tarts.

Apple trees are wonderfully useful. There are so many varieties, including the dwarf versions that are suitable for pots, and there is little more satisfying than picking an apple off your own tree and munching it as you stroll around your garden.

If you grow them in tubs, you can plant other crops around the base, such as strawberries, lettuce and beetroot – basically anything that doesn't have a deep root.

SOIL/SITE

Apple trees need plenty of room for their roots, so when planting make sure the area is big enough, or the tub is as large as possible.

PLANTING

Plant between late autumn and early spring. Plant deep and make sure the roots do not dry out at all during the first year. This is especially important for a container tree.

TREATMENT

Keep moist but not wet. Mulch through summer. Prune in winter when the tree is dormant. Always prune to just above a bud and, if limiting height, prune the 'leader' (the single upright branch that is reaching for the sky) right back.

AFTERCARE

Watch out for scab and insect damage. You can safely use Bordeaux mixture on your crop but it inhibits ripening for a short time and so I tend not to use it.

HARVESTING AND STORAGE

Pick when ripe. Store wrapped separately in newspaper in a single layer in a cool, dry cupboard or shelf – if you use only completely undamaged fruit and leave the stalk in they should keep for three to four months, depending on the variety.

Apples store well in the freezer if peeled and cut ready for cooking. They will go brown easily, so either use a drop of lemon juice to help retain colour or work fast and freeze immediately.

EXPECTED YIELD

Dependant on size and maturity of the tree.

VARIETIES

Italian varieties will not do well in northern climates so choose something local. I like 'Suntan', which has a lovely fresh, nutty flavour and is good to eat and to cook with.

rocket, apple, walnut and parmesan salad

roquette, mele, noci e parmigiano insalata

This dish is so simple and easy to prepare but tastes divine and, with some top class olive oil and balsamic vinegar, is as sophisticated a dish as any diner could wish for. Walnuts become available fresh in autumn but I always save some of last season's to make sauces and salads in summer. Seasonal apples are available in Italy from late summer onwards.

PER PERSON AS
A LARGE STARTER

1 small apple

4 handfuls of rocket

extra virgin olive oil

lemon juice

rock salt and freshly ground black
pepper

20g (¾oz) top class Parmesan

4 walnuts (8 halves)

I never peel fruit – ours is all organic so I don't need to worry about it. Just give them a wash, cut them in half and de-core. I like cutting them into big chunks so that you can still pick them up and eat them by hand, plus, it's less surface area to go brown.

Place the apple chunks into the bowl with the rocket. The pepperiness of the leaves works so well with the sweetness of the apple. Drizzle with a good extra virgin olive oil just to coat and a small squeeze of lemon juice and season well with rock salt and black pepper. Toss all this together. Shave some Parmesan over, crumble the walnuts on the top and serve.

PRESERVING APPLES

Apples store well if wrapped individually in newspaper and stored in a cool dark place, preferably not touching each other. The phrase 'one bad apple in the barrel can ruin them all' is perfectly true. If one apple goes mouldy and is touching another, then the mould spreads to the next apple, which is why we use the barrier of the newspaper. Apples will only keep for so long before they start to go wrinkly. It depends on the variety but you will be lucky to keep them fresh past Christmas. About this time of year I use up my stored apples rather than let them go mushy. I make chutney (see p.163) with them and freeze the rest in ready peeled and cubed portions. You can also make them into prepared pie filling and then freeze it – this way they will keep their colour better than freezing them without any sugar. Now is also the time to make an apple sauce, which you can store in jars for at least six weeks in the refrigerator.

apple pie
torta di mele alla nonna

This recipe is one I copied after a trip to Alto Adige in the far north of Italy. I loved this recipe because it felt like 'Italian Fusion' food and reminded me that Italy has such a huge diversity of landscape that a hefty pudding such as this, served with custard or cream, is just the ticket when you ski off the Alpine slopes.

SERVES SIX

FOR THE FILLING:

900g (2lb) cooking or robust eating
    apples (not Golden Delicious or
    anything else that goes to mush
    when you cook it)
1 tbsp lemon juice
75g (3oz) unrefined sugar
1 tsp freshly grated nutmeg
40g (1½oz) butter

FOR THE PASTRY:

150g (5oz) butter
225g (8oz) plain flour plus extra
    to dust
pinch of salt
1 tbsp caster sugar
1 egg yolk
2 tbsp chilled water

Make the pastry first: rub the fat in to the flour, salt and sugar. When it resembles breadcrumbs stir in the egg yolk. Add the water and knead until you form a smooth dough. Add a little more water if necessary. Chill for about 20 minutes.

Preheat the oven to 175°C (350°F). Peel the apples if you so wish (I don't because it looks even more rustic with the skins on and you lose a lot of vitamins if you peel), then core and cut into eighths. Put in a large bowl, sprinkle on the lemon juice to keep them from going brown and add the sugar. Add the nutmeg and stir gently to distribute.

Melt the butter in a large pan, add the apples and cook for about 10 minutes on a medium heat, stirring constantly so they don't burn. Take off the heat and put on one side.

On a floured surface, roll the pastry out in a round of about 30cm (12in) diameter to a 5mm (¼in) thickness. Heap the apples high in the centre, leaving a 5cm (2in) border. Gather the pastry border around the apples, folding it over and into the centre but leaving the centre open.

Cook for about 35 minutes until the pastry is crisp. Serve with custard or cream.

cherries

Cherries are grown in Italy a lot. They mature early – in May, normally – and are generally eaten straight away as a wonderful precursor to the abundance of fruit that is to come.

There are dwarf varieties that need less room, but in general cherries like a lot of space and light so that the fruit can mature in the sun.

SOIL/SITE

Cherry trees need plenty of room for their roots, so when planting make sure the area is big enough, or the tub is as large as possible.

PLANTING

Plant between late autumn and early spring. Plant deep and make sure the roots do not dry out at all during the first year. This is especially important for a container tree.

TREATMENT

Keep moist but not wet. Mulch through summer. Prune in early or mid-summer – immediately after fruiting, to avoid infection by airborne fungus, which should not be circulating at this time. Always prune to just above a bud and, if limiting height, prune the 'leader' (the single upright branch that is reaching for the sky) right back.

AFTERCARE

Birds love to pinch your cherries, so if you have a problem, throw a large net over the whole tree.

HARVESTING

Pick when ripe. Cherries store well in the freezer – leave their stone in or take it out, it's up to you. They are great for making jam and also keep well in syrups or as glacé cherries.

VARIETIES

The 'Morello' variety will do well in a sunny, south-facing garden and give those wonderful luscious dark fruits.

cherries in syrup

ciliegie sciroppate

There is no denying that the best way to eat cherries is ripe from the tree but sometimes, if your tree's a big one, you get too many to manage. You can freeze them directly (without even destoning them) or you can preserve them in this old-fashioned way.

MAKES ABOUT
TWO LARGE JARS

1kg (2¼lb) sugar

500ml (18fl oz) water

1kg (2¼lb) ripe cherries, stoned

Place the sugar and water in a large saucepan and bring to the boil. Add the cherries and simmer for about 5 minutes. Remove from the heat and spoon the cherries into sterilized jars. Cover completely with the syrup from the pan and put the lids on tightly. If you are not keeping the cherries for long and can store them in the fridge you can finish here.

However, if you want to create proper storecupboard cherries, which will last for at least 12 months. then go on to:

Put the sealed jars into a large saucepan and cover with water. Bring to the boil and boil for 20 minutes. Leave to cool in the pan. Store in a cool, dark place.

cherry ice cream
gelato di ciliegie

We have a lot of cherry trees on our farm and I often end up with tons of cherries that I need to use up. I make jam, glacé cherries and cherries in alcohol to give as gifts, but the most popular thing I make is cherry ice cream.

Italian ice cream is an art form. One of my favourite sights is the local *carabinieri* (police) parked outside the ice cream parlour, leaning on their squad car and licking huge cones of *gelato* while chatting up the local girls.

I use a normal ice-cream maker for this – they are simple and efficient and worth the small investment if you like ice cream or you have kids.

SERVES FOUR

250g (9oz) fresh cherries, stoned

1 egg

1 cup of milk

1 cup of cream

150g (5oz) sugar

1 tbsp cherry jam

Purée half the cherries and roughly chop the remainder. Beat the egg, then beat in the milk, cream and sugar. 'Cream' the jam and add to the liquid, whisking to disperse. Fold in the cherry purée, stir in the chopped cherries and place in the ice cream maker for 20–30 minutes. Eat immediately or place in a plastic container and store in the freezer

other fruits

figs
fichi

Figs are gorgeous fresh. Nothing could be more Mediterranean. You hardly ever see figs in shops because they don't keep at all once off the tree. All the more reason why you have to have a fig tree in the garden. Suprisingly hardy, the fig tree will produce fruit reliably even in northern climates.

SOIL/SITE
Fig trees are hardy and like bad soil. They particularly need well-drained soil, so grow in a pot or on a south-facing slope.

PLANTING
Plant between late autumn and early spring. Plant deep and make sure the roots do not dry out at all during the first year. Fig trees spread their roots out a lot and need to be contained unless you have loads of room, so they are particularly adapted to containers.

TREATMENT
Let dry out occasionally but not for longer than a few days. Mulch through summer. Don't prune your fig unless absolutely necessary – they don't like it and can sulk for a couple of years afterwards.

AFTERCARE
Birds love to pinch your figs, so if you have a problem, throw a large net over the whole tree.

HARVESTING AND STORAGE

Pick when ripe. They don't store at all, so keep them in the fridge and eat them within a couple of days.

EXPECTED YIELD

Dependant on size and maturity of the tree.

VARIETIES

In its natural habitat, the fig tree is pollinated by the 'fig wasp' but since it only lives in very warm climates you will need to buy a self-pollinating variety. 'Brown Turkey' is one of the most reliable cooler climate figs.

blackcurrants/redcurrants

ribes nero/ribes rosso

Currants are easy to grow and are packed with vitamin C. They grow really well in Italy and to giant sizes, as the fruit likes the sun. Italians refer to these and strawberries as 'frutti di bosco' (shade) because they like a bit of shade in the woods from the relentless Italian sun.

SOIL/SITE

These bush plants will grow well anywhere as long as it's drained. You will get away with planting them on clay soil if they are drained enough. The more sun they get, the sweeter the fruit.

PLANTING

Plant new plants in early spring when the major frosts are over.

TREATMENT

They love a well-fed soil, so manure well. Prune back blackcurrants hard – they only fruit on new branches anyway. Redcurrants are different – prune only when you need to thin a bush.

AFTERCARE

Birds can be a problem – get a cat or cover with netting.

HARVESTING

Pick when ripe. Berries will keep well in the freezer and make great jam.

EXPECTED YIELD

Dependant on size and maturity of the bushes.

VARIETIES

'Titania' is a great variety of blackcurrant to grow in a sheltered spot – its fruits are large and sweet. 'Stanza' is a good one for redcurrants – it flowers late and so is less likely to be influenced by frost.

citrus fruit

agrumi

If you have a conservatory or warm greenhouse, it is a real indulgence to grow an orange or lemon tree. They can be kept quite small, so you can even keep them in the house through winter if you don't have anywhere else to put them.

SOIL/SITE

Citrus trees won't stand more than a very light frost, so keep planted in tubs in order that you can move them around. They need particularly well-drained soil, so grow in a pot or on a south-facing slope.

PLANTING

Plant in a well-drained soil and add plenty of manure every year.

TREATMENT

Let dry out occasionally but not for longer than a few days. Keep in the warmest position in the garden throughout summer and bring inside at the first threat of frost until there is no longer any possibility of freezing weather.

AFTERCARE

Prune to keep their shape or limit size – otherwise leave alone.

HARVESTING AND STORAGE

The fruit is ready in winter. Pick it as soon as it looks ready or leave it on the tree until you are ready to eat it – it stores a lot longer on the branch (2–3 months). Once picked, store in a cool place where air can circulate (not in plastic bags).

EXPECTED YIELD

Dependant on size and maturity of the tree.

VARIETIES

Orange trees are slightly hardier than lemons but it's unlikely that they will survive outside in colder climates. Choose whatever variety you like the look of.

parma ham-wrapped figs with dolcelatte stuffing

prosciutto di parma fichi ripieni con dolcelatte

SERVES FOUR

8 fresh figs

150g (5oz) gorgonzola or dolcelatte cheese (or pecorino if you don't like blue cheese)

4 slices of Parma ham, cut in half

Heat the oven to 190°C (375°F).

Top the figs, removing the stalks. Cut a cross in the top of each fig, cutting about two-thirds of the way down. Squeeze the base of the fig lightly to open it up slightly. Place a square of cheese in the centre of each fig and wrap the ham around the bulb of the fig.

Bake in the oven for about 10 minutes until the cheese is melted. Serve on a green salad as an antipasto.

ricotta tart with berry sauce

torta di ricotta al bosco

This is a traditional recipe using lovely ricotta cheese. It can be eaten chilled in summer or warm in winter. If you want to make the traditional Tuscan *torta della nonna*, simply add 100g (3½oz) toasted pine nuts to the filling and don't cover with the fruit sauce.

SERVES SIX

500g (1lb 2oz) sweet shortcrust pastry

FOR THE FILLING:

350g (12oz) ricotta

150g (5oz) caster sugar

juice of 1 orange

6 eggs, beaten

140ml (4½fl oz) single cream

FOR THE SAUCE:

2 handfuls of berries 100g (3½oz)
    caster sugar

Preheat the oven to 170°C (325°F). (Roll out the pastry, line a 22cm (9in) tart tin and blind bake for 15 minutes.

Mix the filling ingredients together, using a fork or whisk. Pour into the pastry case and cook for a further 25–30 minutes until the filling is set.

Remove from the oven and chill on a plate.

Bruise half the berries and leave the others intact. Place all the berries and sugar in a pan and cook gently for 10 minutes until the fruit and sugar have reduced to a syrup – add a little water if necessary (depends on how juicy your fruit are). Add more sugar if the sauce is still tart.

Chill the sauce and serve with the flan.

drugged lemon chicken

pollo al forno con limone e droghe

This recipe comes from my good friend and excellent cook, Eva. When she mentioned she was 'drugging' the chicken one evening when she was cooking for us I looked at her like she was going mad. It turns out that this is a normal Italian expression meaning 'seasoning with herbs'.

SERVES FOUR

1 large chicken

olive oil

rock salt

1 bunch each of thyme and rosemary

3–4 garlic cloves, peeled

5–6 lemons, cut into wedges

Heat the oven to 220°C (425°F).

Paint the chicken with olive oil and sprinkle with salt. Stuff half the herbs and garlic, and 3 lemon wedges inside the chicken. Place the remainder around the chicken in the roasting tin.

Place in the oven and roast for 20 minutes per 500g (1lb 2oz) chicken, basting the chicken with the oil every 10 minutes or so – add some water to the tin if you think it looks like burning on the bottom.

To check that it's cooked, insert a skewer into the thickest part of the thigh – the juices should run clear.

Serve with rosemary roast potatoes.

Preserving

preserving
your crops

I used to think that there was some sort of mystical method to making preserves and bottling produce. Lurking in my subconscious were such scary words as 'botulism' and, possibly worse after all the hard work you had put in, the phrase 'it went mouldy'.

However, after being sent into one of my neighbour's *cantinas* (store room) to fetch a jar of preserved *melanzane* I became an immediate convert. It was late autumn and the place was packed from cotto-tiled floor to chestnut-beamed ceiling with jars of… stuff. It was like finding a treasure trove. This family were certainly not going to starve during the winter. Monica had preserved just about everything: either pickled, under oil, made into pastes, or dried. I suddenly felt my lack of ability in the preserving department as equal to not having a pension, or not being able to keep my family warm. I asked Monica how she learned to do all this and the reply was the same one I get almost every time from my Italian friends: 'My *Nonna* taught me.' Lacking an Italian *nonna* (grandmother), I asked Monica to help, which she delighted in doing because, quite simply, it was all so easy to do…

- The kitchen gardener very often produces more than it is possible for the family to eat fresh. The answer to this is to either freeze the surplus (although not every vegetable freezes well) or learn the traditional methods of storing foodstuffs for consumption later on in the year.

- In Italy crops such as tomatoes are grown in profusion just so they can be turned into passata to keep the kitchen going through winter.

- It's worth remembering that all preserved fruit and vegetables lose a considerable amount of vitamins during the process and storage so it always makes sense to eat things straight out of the garden if possible, but when you have a glut you need to know what to do with it.

- The secret of all preserves is to kill any dodgy bacteria that might have been living on the fruit or vegetables, to kill any pathogens that might be lurking in the jars and to make sure that the containers are sealed properly so that nothing nasty can get in during storage. Follow these rules and you will be perfectly safe eating your preserved produce.

Chutney is fruit and/or vegetables cooked, flavoured with spices and preserved in vinegar. It is not particularly Italian, there isn't even a word for it in Italian, but I have developed a recipe that should be in every Italian cookbook!

The wonderful thing about chutney is that you can add pretty much whatever you have ripe in the garden in almost any combination and it will taste great.

I use chutney with cheese, meat and to flavour winter stews.

STERILIZING CONTAINERS

You can sterilize jars and their lids by boiling them in a large pan for 15 minutes. Make sure they are sterilized just before using and dry them thoroughly before use with a clean towel. Rcipes use 500g (1lb 2oz) jars.

balsamic garden chutney
giardino chutney balsamico

MAKES ABOUT SIX JARS

1kg (2¼lb) tomatoes

4 medium onions, peeled and finely diced

500g (1lb 2oz) courgettes, finely diced

500g (1lb 2oz) apples, cored and fincely diced

3 garlic cloves, peeled and finely diced

500ml (18fl oz) water

150g (5oz) brown sugar

salt and freshly ground black pepper

1 tsp ground ginger

1 tsp nutmeg

500ml (18fl oz) red wine vinegar

500ml (18fl oz) balsamic vinegar

These ingredients are an example only – with chutney you can put just about anything in.

Sterilize your jars (see above).

Mince the tomatoes (you can peel them if you want but I never bother) and place in a large steel pan (don't use copper or brass – it'll corrode). Add the onions, courgettes, apples and garlic and pour in the water. Cover and cook very slowly, stirring regularly, until all the ingredients are soft. Don't worry if it starts going a bit mushy – this is what you are aiming for. Don't add more water unless it's completely dry and threatening to stick.

When the water is all gone and your vegetables are soft (normally about 20 minutes) add the sugar, seasoning and spices. Then add the vinegar. You need to use about ½ wine to ½ balsamic vinegar and you only need to use sufficient vinegar to just cover the ingredients in the pan.

Simmer gently, stirring continuously at the point the chutney starts to thicken. It is ready when you pull your spoon along the bottom of the pan and you see a trail of metal behind it – ie the water has been reduced off and there is no more spare liquid.

Spoon immediately into the sterilized jars. Make sure there is a space between the lid and the chutney, as the acid in the vinegar can corrode metal lids. Screw on the lid as tightly as you can and leave to cool.

Chutney made like this should store for well over a year.

pickles

sotto aceto

No Italian *antipasti misti* is replete without *giardiniera* – pickled garden vegetables.

MAKES SIX JARS

1 litre (1¾ pints) white wine vinegar
    with some extra just in case

2 bay leaves

2–3 cloves

1 tsp peppercorns

1 tbsp salt

1 small cauliflower

250g (9oz) button onions, peeled and
    soaked in cold water for 1 hour

250g (9oz) carrots, peeled

250g (9oz) white celery, stripped
    of filaments

100g (3½oz) green beans

100g (3½oz) artichoke hearts

100g (3½oz) peppers, seeded

You can use pretty much whatever vegetables you have available but the following work well.

Sterilize your containers and lids (see p.163).

Boil the vinegar with the herbs, spices and salt. While it's heating, separate the cauliflower florets into small pieces, cut the carrots and celery into sticks, cut the beans, artichoke hearts and peppers into bite-sized pieces. When the vinegar comes to a boil, add the vegetables and cook them for about 15 minutes.

Using a slotted spoon, transfer them to the jars and pour the boiling hot vinegar over them; have more boiling vinegar at hand, as sometimes you need more to top up the jars. Make sure that they are not over-filled, as the vinegar can corrode the lids.

Screw the lids on tightly and let them cool. Store them in a cool dark place for a couple of weeks, and they're ready for use. They will keep for at least a year.

mixed vegetables under oil

misto sotto olio

This method of preserving works best for high-acid foods such as tomatoes but you need to be careful with any other type of veggie because botulism can occur in low-acid vegetables, even if they are preserved under oil. The safest thing is to store in the fridge and use within a couple of weeks. Use the oil for cooking.

MAKES TWO JARS

1 bunch each of fresh thyme and
    oregano, minced

1 litre (1¾ pints) olive oil

1 aubergine, cut into thin slices

1 large courgette, cut into thin slices

1 pepper, seeded and cut into
    thin slices

3 sun-blush tomatoes

salt

Sterilize your jars (see p.163). Preheat the grill.

Add the minced herbs to the oil. Using a pastry brush, brush the vegetables lightly with the oil mixture and sprinkle a tiny bit of salt on each.

Grill the vegetables until browned on both sides.

Whilst still hot, pack the sliced vegetables into the jars, adding a little oil after each layer. Seal well and store in the refrigerator.

classic passata

Passata is simply tomato paste. Traditionally it is made in the fields in late summer when most of the tomatoes are ripe. Usually the nonni (grandparents) prepared and cooked the passata in great big pans over a campfire whilst the rest of the family carried on picking.

You can add ingredients to your passata such as herbs or onions and garlic. The only thing you will never find in passata is oil.

Passata rustica is made with the seeds and skin left in. Otherwise you need a tomato mill that grinds the fruit and spits out the seeds and skins separately. Even though I've used one and fed the leftover bits to the chickens, I still feel it's a waste – *passata rustica* tastes exactly the same and has more body to it.

tomatoes (any sort but plum are traditional)
1 tsp salt for each 1kg (2¼lb) of tomatoes

Sterilize your jars (see p.163).

Either mill or dice the tomatoes. Place in a large pan with no lid (at this point add other diced ingredients if so desired). Heat gently to a simmer and cook for 2–3 hours. The time depends on how watery your tomatoes are and how thick and concentrated you like your passata. Stir occasionally and make sure in the later stages that the passata does not stick and burn. Add the salt to taste.

Use a large funnel and soup ladle to fill your jars. Secure the lids tightly and store for up to one year.

cheat's sundried tomatoes
pomodori secchi

Sundrying is a classic Italian way of preserving fruit, including the famous sundried tomatoes.

In cooler climates there is normally not enough sun and too much humidity to sundry. However, there are other ways to dry your surplus crop.

tomatoes
salt
chopped herbs

Put your oven on the lowest setting possible. Halve the tomatoes and remove the pips with a spoon. Place on a rack, inside facing upwards, and sprinkle with a little salt and the herbs to taste.

Place in the oven for about 2 hours or until they look dry.

Once they are cold, store in an airtight container and eat within two weeks, or put under oil (*sotto olio*, see p.164), traditionally with some capers in there too and save for around one month in the fridge.

proverbs *proverbi*

As you would expect, Italian country proverbs are down to earth and full of sensible advice. The following are gleaned from my 'contadini' (peasant farmer) neighbours who love nothing more than to quote incomprehensible phrases at me to test my understanding of the language and the land.

Se piove per San Gorgonio tutto l'ottobre e un demonio

If it rains on San Gorgonio day (9 September) the weather in October is going to be terrible.

Anche la regina ha bisogna della vicina

Even the queen needs her neighbours.

Ottobre piovosi, campo prosperoso

A rainy October means a prosperous field.

Oca, castagna e vino per festiggiar San Martino

Feast on goose, chestnuts and wine for San Martino.
(25 November)

Metter su cresta

'Put on the cockerel's comb' – take charge.

Natale con i tuoi, Pasqua con chi vuoi

Spend Christmas with your family and Easter with whoever you prefer.

L'ultimo vestito ce lo fanno senza tasche
Your last clothing is made without pockets – you can take
nothing with you when you die.

Non avere neanche gli occhi per piangere
So poor, you don't even have eyes to cry with.

Per San Valentino, primavera sta vicino
At Saint Valentine's Day (14 February) spring is very close.

Albero che non fa frutta, taglia taglia

If a person/plant isn't being productive, get rid of them.

Dov' entra il sole non entra il dottore
Where there is sunshine there is little need for doctors.

Giugnio freddo, contadin dolente

A cold June makes pained peasants.

In Luglio e ricca la terra, ma povero il mare

In June there is plenty of produce from the land but
not much from the sea.

Se piove in Agosto piove olio, miele e mosto

If it rains in August it means the oil, honey and grape harvest will
be abundant.

Pensare e molto lontano dall'essere

Thinking it is a long way from doing it.

Avanti Natale ne freddo ne fame

Before Christmas you are never cold and never hungry.

Chi vuole un buon erbaio, semina in Febbraio

If you want a good herb garden, plant the seeds in February.

Aprile poiovoso, Maggio ventoso, anno fruttuoso

Rainy April, windy May, fruitful year.

Maggio ventoso, grano generoso

Windy May, good grain harvest.

Aprile carciofaio, Maggio ciliegiaio

April artichoke harvest, May cherry harvest.

Corvi con corvi non si cavano gli occhi

Crows don't peck other crows' eyes out – there's safety with
people like yourself.

index

PICTURE CREDITS

Front cover: main picture GAP Photos; recipe picture Yuki Sugiura; other images Ian Nolan. *Back cover*: Ian Nolan. *Internal images*: recipe and recipe ingredient photography Yuki Sugiura; atmospheric and location photography Ian Nolan. *Chapter openers*: 16–17, 48–49, 134–135, 158–159 Shutterstock®.

author acknowledgements

I would like to thank everyone at Anova for making this book possible. Also Ian, for making the photography such fun.

The book is dedicated to my friend, Emma Perry.